How To Make Poultry Pay

Trustworthy Money-Making Information Covering The Experience of Many Breeders

by Inland Poultry Journal

with an introduction by Jackson Chambers

This work contains material that was originally published in 1909.

This publication is within the Public Domain.

This edition is reprinted for educational purposes
and in accordance with all applicable Federal Laws.

Introduction Copyright 2017 by Jackson Chambers

Self Reliance Books

Get more historic titles on animal and stock breeding, gardening and old fashioned skills by visiting us at:

http://selfreliancebooks.blogspot.com/

Introduction

I am pleased to present yet another title on Poultry.

The work is in the Public Domain and is re-printed here in accordance with Federal Laws.

As with all reprinted books of this age that are intended to perfectly reproduce the original edition, considerable pains and effort had to be undertaken to correct fading and sometimes outright damage to existing proofs of this title. At times, this task is quite monumental, requiring an almost total "rebuilding" of some pages from digital proofs of multiple copies. Despite this, imperfections still sometimes exist in the final proof and may detract from the visual appearance of the text.

I hope you enjoy reading this book as much as I enjoyed making it available to readers again.

Jackson Chambers

THE breeding of domestic poultry has reached a point in this country where the most down-to-date methods must be applied if the results sought for are to be attained. Every year there are thousands of new recruits added to the rank of poultry enthusiasts. They come from all walks of life and are interested alike in making a success of their venture and are looking for the most reliable poultry information that can be found that will teach them how to start right.

Realizing the needs of this great army of amateurs for reliable poultry information we have compiled what we believe to be the most authentic work on how to make poultry pay that has ever been issued. In this book we have tried to give the cold, plain facts in reference to poultry culture—facts that can be depended upon. We have not attempted to paint rosy pictures of the great imaginary possibilities of the business, but have given the experiences of men who have made a success of poultry culture, and by following this advice we feel that we can teach others how to succeed and make money out of the business.

In our brief history of the many popular breeds we have attempted in as few words as possible to give a concise and authentic report of their origin, with such facts about each breed and variety as will enable the beginner to make the best selection for his foundation stock.

In fact, we have tried to give to the readers what the title of the book indicates—the information necessary to make poultry pay. If we succeed in doing this we shall feel that our time and labor have been well spent.

THE PUBLISHERS.

PARTRIDGE WYANDOTTES.

Second cock and first hen at the World's Fair Poultry Show. Owned and exhibited by W. A. Doolittle, Sabetha, Kan.

THE BREEDING STOCK.

Selecting and Caring for the Breeders—Selecting Breeding Males, Yarding, Housing and Feeding.

By THEO. HEWES, Editor Inland Poultry Journal.

TO produce strong, healthy chicks it is important that we know the parent stock. We must know them as individuals and something of their ancestors; without this information we are taking long chances—in fact, chances that we cannot afford to take if we expect satisfactory results. Delicate, poorly matured specimens cannot reproduce strong, healthy offspring. This is true of all animal life, and especially true in poultry culture. At no time in the breeding of poultry, either fancy or for commercial uses, is there a more important task before the breeder than the proper selection of the parent stock.

Next to the selection of the breeders is the conditioning of the birds so that the best results may be obtained.

we admit it or not, it is a fact just the same our fads have much to do with our success in life. Fortunately, the lovers of poultry have every opportunity to gratify their taste, for all breeds of standard fowls can be made to pay a nice profit on the investment and one has but little advantage over the other, and while the heavier weight breeds are more valuable for market poultry they require a longer time and more feed to produce them. The smaller breeds will grow to maturity quicker than the larger ones, and as a rule will begin laying from five to eight weeks earlier than those of heavier weight, but when we take a twelve months' average, considering market value as well as egg yield, we will find there is but little if any advantage one over the other. So in starting in

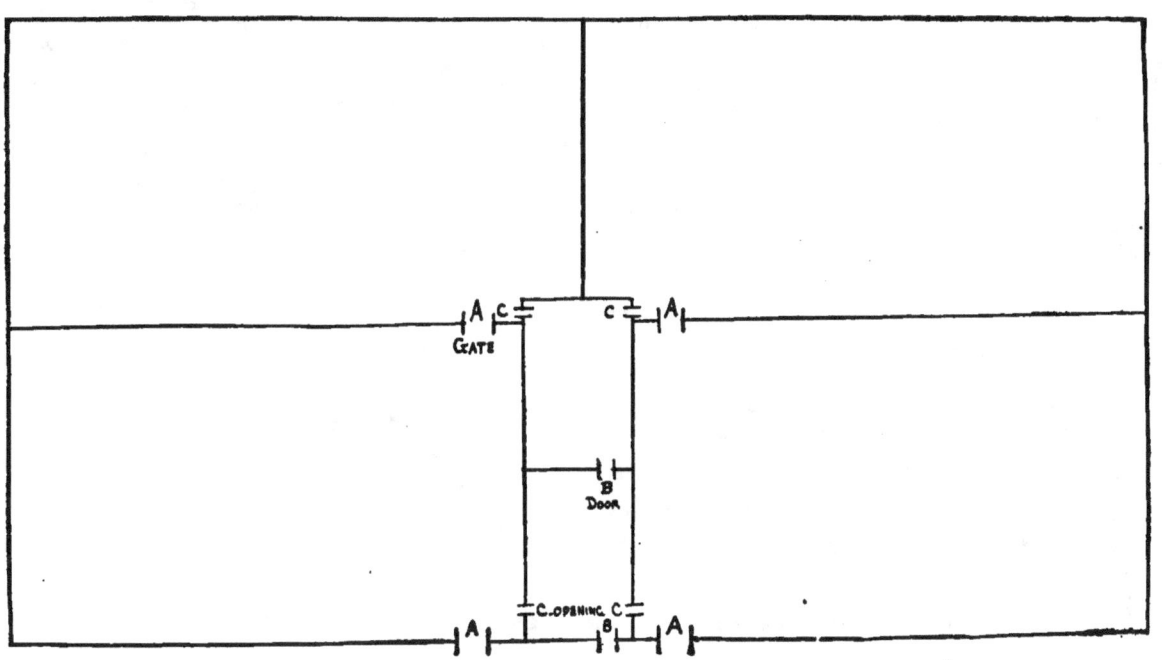

FIG. 1. SHOWING PLAN FOR LAYING OUT BREEDING YARDS.

There is no great secret about poultry culture; there is, however, a "know how" that can be learned by experience, and we will try to save you this experience by asking you to avoid the pitfalls that we were unfortunate enough to fall into, and to profit by what we have learned to be good.

It is our aim to give in this book the actual experience of men who have made a success of poultry culture and have no other source of revenue than that derived from the poultry business.

In selecting breeding stock, whether for fancy or commercial breeding, always use thoroughbred fowls. It doesn't pay to use the common or scrub stock. This is true of all live stock breeding, but in none of them more important than breeding poultry. By starting in this way you will get a better average egg yield; you will get better results in quick growth of chicks; more weight for the same amount of food and above all else better prices for the surplus stock, whether sold to the market men or the poultrymen for improving their stock.

The breed or variety that will do best in your hands depends largely on your own likes and dislikes. Our experience has been that one always does best with the breed they most admire. This rule will hold good in every walk of life. That which we think most of will receive our most careful attention, while something we do not admire will receive only such care as we are compelled to bestow upon it. We are all human, and, whether

the business it is safe to select whatever breed you most admire and feel that you are making no mistake.

In the following pages we give you a few chapters from our own experience which we trust will prove of value to you.

After you have selected the breed you like and have found a suitable place for the breeding of fowls, select your breeders from stock that has not been subject to roup or other diseases; select birds as near standard size as you can find, those not over large nor very small. The nearer you can come to standard size and weight the better for your foundation. Should you desire to increase the size of your birds then select large females, as this sex has most to do with the size of the offspring, while the shape is more affected by the male than the female. These are points well worth remembering and are called attention to at this time as we often see breeders, especially those just starting into business, select a large male to breed on to small hens, expecting great results only to be disappointed; while a few large hens mated to a medium sized male would have given the object sought. Our experience with large over-weight hens has been far from satisfactory. As a rule they are poor layers and the percentage of fertile eggs from them is below the average, while small hens as a rule will outlay the larger ones and their eggs are strongly fertilized, so many of them lay small eggs and a great many of the eggs are bad shaped, while the standard weight hens almost invariably are our best layers, not alone in number but uniform size, and the eggs invariably hatch well both under hens and in incubators and give us our strongest chicks. For this reason we urge upon new breeders the adoption of standard weight fowls rather than the larger or smaller.

The American Poultry Association has given the matter of weight careful consideration, and the Standard adopted by them has proven highly satisfactory to the best informed breeders in all sections of the country, and all poultrymen, whether breeding for fancy or market, should have a copy of the Standard, as it describes in detail every breed of land and waterfowls classed as a thoroughbred.

Selecting Breeding Males.

In selecting the male bird to head a pen always bear in mind that he is half the pen from a breeding standpoint. No matter how many females you may mate with him every chick that is hatched from the pen is sired by the one male, and for this reason the best judgment must be used in the selection.

As we have before stated the male has much to do with the shape of the offspring and we will add the stamina of the chicks depends largely on the vigor of the male. The male bird to be a good breeder, whether cock or cockerel, must be typical of the breed he represents. He should be up to or a trifle over standard weight, and this weight should be attained by natural growth and not by overfeeding. He should be active, alert, ever on the lookout and one of the first to notice an intruder. Care should be given to see that all females have the same attention from him, as we know of many cases where the male bird would make favorites of certain females in his harem and pay no attention to others. Should this occur mate or confine another male in the adjoining pen so that the breeding bird can see him and you will note a decided change in his conduct.

Don't Overmate.

It is best to undermate rather than over-mate, as your success with strong chicks depends on this as much as any one thing. If germs seem weak then reduce the number of females; on the other hand, if male bird is giving too much attention to the females add two or three more hens to the pen. The number of females to mate with one male depends some on the variety and more on the amount of room you have in your breeding yards. We usually mate from eight to eleven females to one male in the middleweight breeds, such as Plymouth Rocks, Wyandottes, Rhode Island Reds and breeds of that size, while in Leghorns quite often one male will care for fifteen or twenty females, while the Asiatic breeders find themselves compelled to reduce the number to from three to eight, depending entirely upon the vigor of the male bird at the head of the yard.

Plenty of Room is Important.

The vigor of both the male and female seems to be improved by ample range, and for this reason we advise making the breeding yard as large as space will allow. We have found that a yard of one-fourth acre will furnish green stuff for a pen of twelve birds, that is, it will replenish as fast as used, but we have had better results where the same number of birds had unlimited range.

Where fowls are confined in breeding yards we have found by giving two runs to each pen that better results followed than by using the single yard. Under this system we would use one yard at a time, allowing the fowls to run in one for perhaps ten days, then change them to the other for the same length of time. This can be done at no great expense by using the double yard system.

In Fig. 1 we show a house and runs suitable for two pens of breeding birds. The house is 10x20 with partition in middle. All space on inside is used, all doors opening directly into the house, one door opening at end and one in partition. There is a double sash window in each room that slides back on sills against the wall, allowing of free ventilation in summer. This house is 9 feet high in front and 5 feet at rear. Sides, ends, partitions and doors are all made of barn siding dressed on both sides. Framing lumber 2x4 scantlings for sills, frames and rafters. The house is covered with best quality of cedar shingles. There are three posts set 26 inches in the ground, one on inside end and one in center. They are nailed to frame at end and to center rafter at partition, and in this way will brace the house against the strongest wind that blows. We use no floor, but fill up with earth to the top of the sills. This will keep your house dry, no matter how damp the weather may be outside, as the dirt on the inside being higher than the surface, moisture will not rise above the level. In winter the house is lined with tarred paper on the inside and cracks are battened with 2-inch strips on the outside. We have kept single comb fowls in houses of this kind when the outside temperature was 12 below zero without having a comb frosted. In winter one-half

FIG. 2. SHOWING INTERIOR ARRANGEMENT OF POULTRY HOUSE.

bale of straw is used in each room every ten days. This is laid in center of pen with wire cut but not spread out; we then throw corn, wheat or oats in the straw and let the fowls scatter the litter in order to find it.

In constructing a house of this kind, where one does the work himself, it can be built for $50 in the Central or Western States, depending some, of course, on the price of lumber, and a house of this dimension will accommodate two breeding pens in summer and will house thirty-five females to each room during the cold months of winter. By using barn siding 14 feet long there will not be an inch of waste material. The five feet sawed off of your nine-foot front makes your back wall, and in making ends and partitions the short board sawed from the front, by reversing end for end, will just match the lower side, and by working from front to back, the boards being reversed each time match completely, and by using lumber just twelve inches wide you will meet at the center on both ends and partitions without a half inch of open space.

In Fig. 2 is shown interior of same house, showing arrangements of dropping boards, perches and nest boxes. Here again you will note that the 14-foot lumber is used without waste. Each board is sawed into three pieces, each 4 feet 8 inches long. By nailing them to three cleats, one at either end and one in the middle, you have a dropping board 4 feet 8 inches long by 3 feet in width. We then use three 6-inch boards 3 feet long and nail to them two strips of the same material that was used for the cleating. They are nailed eight inches from either end and we have two perches 4 feet 8 inches long removable

FIG. 3. BROOD COOP AS USED BY G. R. HASWELL, CIRCLEVILLE, OHIO.

and so arranged on the dropping board that the droppings are caught. The dropping board is set on a strip nailed to the wall at one end and wired to the rafter at the other so that no post is required. Roosts and dropping boards can be removed from the house for cleaning. Ordinary cracker boxes with the lids on are used for nests with holes sawed in end of box with open end next to wall; in this way the hen is out of sight when laying and by using a small hinge door in back of box the eggs can be gathered without moving the boxes.

How and What to Feed Breeding Stock.

There have been hundreds of pages written along this line. The advice of our best breeders has been sought by editors and all kinds of suggestions made as to the proper kind and the amount of food in order to get the best results.

In looking over the carefully prepared articles by some of our so-called experts, we take the advice of one and are about to give it a trial when another letter from apparently as well informed writer contradicts the first and still another apparently as well informed contradicts them both. We cannot say that the advice of these experts is all bad, neither can we say it is all good. The conditions surrounding the three breeders may have been so vastly different that all were right so far as their locality was concerned. What is good on the Atlantic coast may not do on the Pacific coast or vice versa. So what we have to say in regard to feeding breeding stock is based entirely on our own experience and has proven the most satisfactory in our own location in the Middle States.

We have made it a point to feed our fowls at the least possible expense, not miserly by any means, but to give such food as would give best results at the least possible cost, and we have never found anything that would surpass pure wholesome farm raised feed properly prepared. Corn, wheat, oats and millet for grain with beets, cabbage, onions and clover tops for green stuff with an occasional feed of green cut bone. With these feeds we will take our chances as to growth conditions and egg yield with any so-called balanced ration ever prepared.

Our best results in feeding millet is when fowls are confined in winter quarters; we then feed millet hay and allow the fowls to thresh out the seed. When corn, oats or wheat is fed whole, the best results come from feeding it in the morning, scattering it in litter and allowing them to work it out. For mash the one we have ever found good is one-third bran, one-third ground oats, one-sixth ground corn and one-sixth clover tops. This should be mixed well together and fed in the evening just before fowls go to roost. This mash should be prepared in the morning by mixing well with boiling water and allowed to set until evening. In cold weather add about one gallon of boiling water to ten gallons of feed just before using. This will warm it up and fill the crop with warm, healthy food during the night and bring eggs in plenty. Cabbage, beets and onions can be fed raw, but the better way is to boil them to a pulp, using a large vessel and plenty of water. When thoroughly cooked thicken by using pure wheat bran or ground oats. About one feed each week with the vegetables will be found sufficient and let this feed take the place of the mash at night. Understand during the winter months we feed but twice a day—morning and evening. Green bone should be used twice each week

A VERY SATISFACTORY EGG TESTER.

and should be fed in the middle of the day. Give them only about one tablespoonful to each fowl and you will have better results than where too much is used. There is no one feed that will do laying or breeding birds more good when confined than green cut bone, and no poultry plant is quite complete without a bone mill. Grit should be kept before fowls at all times, especially when confined, and I have never found anything in this line that surpassed the Mica Crystal grit. It can be purchased from the poultry supply or incubator men at the average cost of $1.00 per hundred pounds.

We believe we have given to the amateur all the information that is necessary for a successful start in poultry breeding and if our advice is followed and the proper care and judgment used by the operator we are confident that success will follow.

Selecting and Testing Eggs.

Eggs that are used for hatching should be carefully selected. As to age of eggs unfitting them for incubating years give you 85 per cent. of whatever color you desire.

During incubation, whether by hens or artificially, test the eggs between the fifth and seventh day. This can be done best at night, and is a simple process. There are any number of testers on the market, good, bad and indifferent, but we have never found one better adapted for the use of a beginner than the one illustrated on this page, which can be purchased from any incubator concern or poultry supply house at a cost of about 35 cents

WHITE ORPINGTON PULLETS ON THE FARM OF KNOWLES, YOUNG & CO., NORTH ADAMS, MICH.

purposes there is a wide difference of opinion among breeders supposed to be well informed. In our own experience we find that an egg from ten to twenty days old, that has been kept in a dry, moderately warm room, hatches better and gives us a stronger chick than those set within two or three days after they are laid. We have also noticed that eggs shipped, when they were carefully packed, have given our patrons as good percentage of hatches as those incubated at home. One thing we do advise is the turning of eggs every two or three days that are intended for hatching, as an egg lying in one position too long will settle to one side and sometimes the yolk will adhere to the shell. While we would advise amateur breeders to set the eggs within fifteen days from the time they are laid, at least make this a general rule. We would not advise the discarding of eggs that chance to be thirty days old, especially if they were of great value.

When selecting eggs a far more important matter confronts the amateur than the age. I refer to the shape and general texture of the shell. Select eggs only of good size and general shape, not extremely large but a good average. Be carful to select those with smooth surface; discard the ill-shaped, thin-shelled ones and avoid those with wide pores or the ones covered with specks that resemble lime. If you are catering to the fancy or commercial egg market then determine from your choice buyers the color of shell most in demand and select eggs for hatching of that shade. You will be surprised how soon you can build up a strain of fowls that will lay the same color of eggs. The selection of proper color will in two

and will last for years. In using this tester take any ordinary coal oil lamp, remove the chimney and place the tester on as a flue, and by holding the egg in front of the opening one can tell at a glance whether the egg is fertile or clear. It is well to make a second test about the fourteenth day, as some germs will start and die in a few days from starting, and by making a second test you can discard all but those that are strongly fertilized, thus giving more room in machines or under hens. If you are using hens for incubating and can so arrange, it is best to set about five hens at one time. At the end of the fifth day you will find enough unfertile eggs to allow you to distribute the eggs under four, allowing one to be set over. In this way you are keeping your hens up to their full capacity and will have one less hen to handle at hatching time. Where good coops are provided two hens ordinarily will hover all the chicks that the four will hatch and if you so desire you can re-set a part of those that were sitting the first time. Six weeks in ordinary weather, where sitting hens are well cared for, will not injure them at all.

In Fig. 3 are shown some brood coops used for outdoor purposes that have given entire satisfaction in the yards of the writer, also in that of several other breeders that have given them a trial. The photo was forwarded to us by Mr. G. R. Haswell, Circleville, Ohio. He claims it has been the most successful coop for raising young chicks outdoors that he has ever found. These can be constructed at a very small expense.

WHITE PLYMOUTH ROCKS.
Bred and owned by John Landis, Edinburg, Ind.

MARKET POULTRY RAISING *

A Long and Complete Article Giving Advice About Those Things That Must Be Considered in Establishing and Successfully Conducting a Practical Poultry Farm.

By D. W. INGERSOLL, Shermerville, Ill.

THAT money can be made in the market poultry business no one can deny, but that a large majority of those giving up more or less lucrative positions to attempt the poultry business fail to succeed is unfortunately true. The cause of most failures is not far to seek—one word, inexperience, which explains the majority of them. There are, of course, other reasons for failure, but nothing so vital as lack of experience.

Locate with an Eye to a Future Market.

The most important part of the poultry business is the market you hope to find for your products, but unfortunately it is the last factor considered in the problem where it should be the first. It is not necessary to go out and engage customers before you have anything to sell, but it is essential to ascertain whether you can dispose of your output without throwing it on the open market. I do not believe it possible to succeed with market poultry unless you can obtain a better price than is offered in the market, where farm-raised stock competes. It may be that with a large capital so that time could be used in building up a reputation for excellence, a market poultry plant could afford to offer its products at a slight premium above the market prices on the open market, but I believe the margin of profit to be so small as to be unsafe in most cases. It is far better to investigate the local conditions where you contemplate selling; ascertain who the people are that live on the best the market affords, and when you have something to sell go to them and show something better than they have been able to secure. Such a course will give you a demand that will increase much faster than your ability to supply it, and at prices that include the commissions as well as the middlemen's profit.

Before you sell stock you must install your plant and raise the birds. So let us discuss the location. First, one fact is evident—it must be sufficiently near your market so that time and excessive express charges may be saved. The second consideration is soil, elevation and slope, of which slope is the most important. A good pitch to the south or southeast is an advantage, though none of these things are absolutely essential; they merely facilitate the work of keeping the plant in good condition. A high, sandy ridge sloping south is an ideal location; stiff, clayey soil lying low and draining north is the poorest, but even that with proper care will allow successful chicken rearing, its chief fault being the foul surface accumulations, which the surface water constantly washes up to the house, making necessary frequent working of the ground and as plowing is often impossible on account of fences and shrubbery, the turning of such soil must be done by spading—a laborious task.

The Houses.

After a good many years of experience I am convinced that in this climate open front scratching sheds in some form are essential to the health and, consequently, to egg production and vitality in the progeny. I have personally built a number of chicken houses of different types and have finally settled to my own satisfaction the most satisfactory house from all standpoints, location considered. Where the location permits permanent houses I build double houses, scratching sheds at each end and roosting and laying in the center, of the following dimensions: over all, 48x12, rear wall 5 feet, front 6 feet 6 inches, roosting room 9 feet 6 inches by 11 feet, double walled, floored and ceiled. Window 3x4 and door 2 feet 6 inches by 6 feet in each room. Partition between rooms is single thickness, outside walls are backed with heavy building felt on both sides of studding before being sheathed. Scratching sheds are each practically 14x12, with 2-inch mesh netting fronts and drop curtain; no floor. Roof is covered with 3-ply roofing paper. These houses are set on charred cedar posts, set 2 feet 6 inches in the ground and 4 to 6 inches above ground, 6 feet apart. Scratching sheds have a base board on inside of posts and are filled with layers of gravel and clay to insure dryness, on top of which is litter a foot or so deep, which covers the entire floor. Such a house will accommodate thirty Plymouth Rocks or forty Leghorns on each side.

For movable houses I find 8 feet by 16 feet a convenient size, using 2x6 studding on edge for runners or sills. A good team will easily drag such a house about and by dividing the house into a roosting room 6x8 feet and shed 10x8 feet and building them in pairs you can concentrate them in winter and save labor. Then when spring comes separate them again and you get the best results, for your flock should be scattered as widely as your land permits without increasing the work of caring for the stock. It is an established axiom that the smaller the flock the better the results.

These houses are for the layers and breeders and farther on I will give a description of a house I have evolved that comes the nearest to overcoming chicken house faults economically than anything I have seen. In addition to your breeding houses you will have a number of portable colony houses. I build these 3x8, 3 feet high in rear and 4 feet in front, a single light of glass 10x14 inches in each end and the roof hinged; a board floor and two 8-foot perches completes them. In addition to these any large box can be used to advantage when roofed and put on runners.

About Brooders and Incubators.

For rearing the chicks to a colonizing age a sufficient battery of outdoor brooders is required, and whatever make you purchase be sure that of three points. First and most essential a constant stream of heated fresh air should flow under the hover, and you need not expect it to enter through fringes or curtains. Second, the top should lift to permit cleaning and a purifying sunbath. Third, the heater, of whatever sort, should be sufficiently accessible to permit thorough cleaning when the lamp or oil stove smokes, as it surely will some day. If you have a lot of flues with elbows a single smoking puts it in the trash heap, for unless the soot is thoroughly removed it will smoke at a moment's lack of notice. Simple as it seems there are very few brooders on the market that conform to these essentials, and that they are essentials any one with experience knows.

The same points are necessary for whatever system of indoor brooding you use. The most convenient plan is a hot-water piping system, divided into sections, but it is exceedingly difficult to apply heated fresh air with this system, and I have never had good results with chickens under than, but ducks seem exactly suited and thrive wonderfully, perhaps, because they are after the first three days much hardier than chicks and as they require brood hovers so short a time the lack of fresh air is not so vital with them. Ducks frequently abandon the hovers when one week old, while chicks require from three to six weeks depending on the season and outside temperature.

For the winter work I use individual brooders for chicks and wherever house room allows it the outdoor brooders are moved in as a three compartment (so-called) brooder is an advantage in doors as well as out.

As for incubators it depends more on the operator than the make of machine, but in selecting one be sure that it regulates fairly and ventilates surely. As for a

---

* This is a grouping of several articles more or less disconnected that appeared serially in the Inland Poultry Journal.

smoking lamp the best machines are the least possible to clean. I say this with limited knowledge, however, as I have used only eight different makes out of a possible hundred. At present I am using both hot air and hot water incubators, but the heated season will see the hot water machines laid up till fall, as the regulation is difficult during hot weather.

I want to say a word here about operating incubators. Many a beginner, finding a machine gone wrong and registering 85 degrees or 110 degrees, throws out the eggs and begins again. Don't do that; it is surprising what variations and extremes of temperature a lot of eggs will go through and hatch fairly well after all, and another thing, follow the manufacturer's directions until experience shows where you can improve on them. Because John Jones writes from Lonesomehurst that he hatched 112 chicks out of 115 eggs and his testimonial is cheerfully printed in the catalogue by the manufacturer, do not think that you must do better.

Beginners fail to realize that 95 per cent. hatches are rather rare. I use incubators now and began their use in 1889. I have painfully acquired some of the rudiments of the art, but I firmly believe that I can select ten hens and hatch a better percentage than any incubator in existence. Then why do I use them? Because the expense of hatching with hens, if you must purchase feed and your time has any value, is prohibitory for the market poultry. I incubated last year nearly 4,000 eggs and all but a couple of hundred from my own yards. That would have been practically impossible with hens. The brooder, however,

I had been reading with great interest the articles of a recent poultry writer, because it seemed that he knew whereof he spoke (which is rare) until he declared over his signature that 2-pound broilers could be produced in eight weeks or less. After that I lost interest in him. The hardest tasks for the beginner with poultry is to realize that the pages of some poultry papers are open to all sorts of theorists who mean well, but whose advice is unsafe to follow, you must learn to pick out practical from a mass of dead trash. Because John Jones in a dream raised four hundred chicks, losing only three, and all because he gave each a little corn after each meal is no reason why you should attempt it.

And right here I want to emphasize the necessity of so arranging your buildings that a terrier can absolutely prevent the lodgment of rats. Rats are an insatiable enemy to chicken raisers, and if an opportunity is given them they take, not what they can use, but all that they can find. On one occasion a rat made an entry into a small brooder house, and in a single day cut the throats and hid the bodies of sixty-three three weeks' old chicks. The next day I bought a thoroughbred fox terrier and now I have a number of them about the place, and find them indispensable. They are encouraged to follow me about all day to the numerous outdoor brooders, brooder houses, colony houses and coops, and if a rat has been or is about I am immediately notified and we get him promptly before he has had time to levy tribute on my flocks. If you have on your place old barns or other places that you can not clear of rats you can at least see that your chick fences

White Leghorn Poultry Yards, Waterville, N. Y.

has the hen beaten to a standstill; no lot of hens can equal a good brooder decently cared for. Credit must be given the incubator for its readiness when sitting hens are scarce, though a good deal can be done to hasten broodiness; for instance, letting eggs accumulate in the nests will frequently hasten it.

Broilers or Eggs.

So much for the discussion of buildings and equipment. Now what shall we raise to make the greatest profit? Naturally we say broilers, but with a reservation. Your private trade will take broilers readily at any season, and my experience has been that there is a season when broiler-hatching is more likely to show a loss than a profit. From December 1st to about February 10th my incubators are idle, because eggs are scarce and expensive, while their frtility is as its low ebb. From October to April 1st I received 40 cents per dozen for dated breakfast eggs, and 30 cents from April to October. Now 40 cents a dozen yields a better and surer profit for me than 60 cents each for broilers hatched from those eggs. Perhaps it should not, but with the high mortality, the cost of fuel, feed and care are likely to show on the wrong side of the ledger and right here I want to say that people who declare over their signatures that eight weeks of any variety of care will produce 2-pound broilers are very likely to be suspected of having an "axe to grind" by any one who has tried it, for it can not be done. Note the experimental station's work on the subject.

prevents the near approach of the chicks, as rats will lie in wait and sally out and catch each chick that ventures too near.

One of the most costly mistakes I have made in the chicken business had to do with the arrangement of my fences. My breeding houses (the permanent ones) are scattered about with an eye to the availability of the site only for the breeders, and when in the early spring I move out my out-door brooders the trouble begins. The fences of my breeding yards are of two-inch mesh, some with a foot wide base board and some without, and the little chicks will pass either very easily in a short time, and when they do, in some mysterious way they pick up lice, and lice in a brooder are a fruitful source of loss. Had I been far-seeing enough to divide my plant into three permanent sections, with absolutely chicken-proof fences I should have been saved much loss. One section for adult breeders, another for colonizing stock and the third for the outdoor brooders and brooder house chicks. It is a hard matter to produce breeders of a rugged vitality if they have been penned up as youngsters in comparatively small yards attached to brooder houses. I have one lot of sectional fence that is built by attaching one-inch mesh netting to sixteen-foot boards, using an upright at each end and in the middle. Each section has a hook at one end and in the middle and an eye at the other, and a fence can be placed of any desired size and shape in a very few minutes and removed just as easily. It is hardly necessary to state that these boards are set on edge which gives a three-foot fence. In addition to these fences I have a few box runs, that is, a frame of about 3 by 6 feet and 2 feet deep, entirely enclosed with netting except floor and one opening in the end that corresponds to the trap door in brooder. Of course there are exceptions, but ordinarily the chickens stay in the incubators from forty-eight to seventy-two hours, and are then divided among brooders heated to receive them. If a three compartment brooder, the second day they are allowed in the sun-parlor or exercise room, and if the weather permits, the third day finds them running about in the box run. They are confined to this a week or perhaps until the sod shows signs of wear. It is then removed and a sectional fence placed to enclose a few bushes and the brooder or some shade provided at least. The sectional fence can be removed when the chickens are growing rapidly and have learned to answer some call for food, so that in case of some sudden storm they may be made snug. I used for some time to imitate the hen's cry of a hawk until I found that while the majority fled to the brooders for shelter, some of the wildest would shelter in the nearest possible spot and lie so close that search often failed to dislodge them. It is inevitable that some of the summer storms should cause loss, but as an off-set the survivors will mature into breeders of strong vitality that he can stand forcing for egg-production without loss of fertility in the eggs or stamina in the progeny. One of the most serious faults with the brooder production of breeders as ordinarily practised is that it violates the essential law of keeping up the standard of any stock. The survival of the fittest. With a hen only the strongest, most rugged chicks survive, while a brooder will raise lots of chicks that are physically degenerate. This accounts in some measure for the trouble amateurs so frequently experience in producing stock year after year that easily reaches standard weight. Of course, this applies to breeding stock only. Broilers I give only enough range to provide green food and exercise. They are forced from start to finish and are unfit for anything but food. I said above that with the hen only the strongest survive; quite frequently those surviving are so dwarfed and stunted by bad care or neglect as to be worthless for breeding also.

I do not believe that any hen is absolutely free from lice. Like David Harum I believe "a few fleas are good for a dog; keep him interested." The constant trouble is to keep the hen from being uncomfortably interested. To that end I use once a week commercial lice killer on the prches. The dropping platforms are cleaned twice a week and an ample dust bath provided. Tobacco dust is used freely in all nests and twice a year hot whitewash with crude carbolic acid is used inside and outside of houses and fixtures. Now, when you set a hen, plenty of tobacco dust will keep her and the nest free from vermin. But when you coop her with a clutch of chicks the weather will frequently prevent, for days at a time, any opportunity for a dust bath, and then lice that have managed to live through the "late unpleasantness" begin to take notice. It is safe to say that a bald chick is a lousy chick, and the time to prevent is passed. The best thing I have tried is a dope made by working a tablespoonful of powdered sulphur into a fourth of a pound of lard, with a teaspoonful of kerosene and ten drops of carbolic acid added. When this is thoroughly mixed a daub on the head and under each wing will relieve the chick without bereaving you.

If you want to use a liquid lice killer it is best to take some slack lime and stay right with the chick with your finger on his pulse so that the relieving process won't go too far. Don't, I beg of you, do as I did once upon a time, paint the under side of a low head room hover, surrounded by curtains with lice killer. In the middle of the night it occurred to me that such a plan was an experiment and that experiments should be watched. It was so warm that I wandered out to the brooder thinly clad, and it was not until I was extremely busy fanning chicks in a vain attempt to restore life that I discovered how plentiful the mosquitoes were. I have never seen advocated the painting with a good house or floor paint of the inside woodwork of brooders, and yet that is one of the simplest and best preparatory measures for success in rearing chicks. Obtain your brooders far enough ahead of time required for use to permit the painting with two good coats of paint. Doing so seals up each crack or crevice that could harbor lice or mites, reduces warping and cracking and, best of all, prevents the absorption by the wood floor of all impurities. A painted floor can be scrubbed very quickly and dries at once, where a bare floor is very slow to dry. Of course, this has nothing to do with the litter in brooders that is essential with any floor, and sand, while cleanly is a very poor litter for little chicks after they are a few days old. I find that the refuse left in the hay-loft after clover and timothy have been removed makes the very best possible litter. It is dust, dry and an inch or more in depth is not beyond the strength of little chicks. Then it is full of hayseed and cloverseed, heads and leaves, which is ideal food to start chicks on in connection with other things. Since using this chaff I have found little need of charcoal as a regular article of diet. Of course, if you feed too bountifully your chicks will eat only the most attractive part of the ration and leave the rest. Where there is much coarse fiber in the refuse I screen it with a quarter-inch mesh to remove most of the stalks. I do not only use this for litter, but in winter I procure this trash baled and use as high as forty per cent. in mash with gratifying results. It is practically within itself a balanced ration and you have only to mix the balance of the mash accordingly. My greatest trouble in feeding has been to find something with great bulk for low food contents without indigestibility and that fills it perfectly. I have never had a crop-bound chicken since I began its use and it is equally good for ducks.

I want to say here that if I am tempted to advance any theories regarding the poultry business that I will label them as such. This is simply a plain statement of what I know to be practical. I have had to advance very slowly at times, but that necessity has saved me many costly errors. One rather costly error in money and time was due to an elaborate article on poultry and fruit raising combined, and I did not discover the error until too late to correct it in most cases. If you plant fruit in your poultry yard you are providing shade and future lunches for your fowls, but disappointment for yourself for some time. The first difficulty comes in protecting the mulching about the young trees from being scratched away by the fowls. After the trees are established the first light crops are utilized by them. I have not yet attained that enjoyable point where a surplus exists after their appetites are satisfied. I am told that the time will come and I try to believe it. That is theory. As for grapes and currants, etc, you will never have a chance to see them ripen if your chickens are like mine. I was compelled to fence off my vineyard thereby badly restricting the yards of one permanent house. The currants and raspberries I moved bodily and have been well paid for so doing. That is an example of the loss that an illy considered article may cause one.

A year ago in one of the poultry papers appeared a carefully drawn plan of a combination farm. In conversation with the editor I said he should comment on such articles in order to prevent his subscribers from injury to pocket and confidence, but he replied that such a criticism would prevent the future offering of much free copy. That seems to be the best reason for criticism. If guarded editorial comment pointed out the fallacy of or experimental nature of these enthusiastic schemes it would save a great deal of misdirected energy in the poultry business

It occurs to me that a brief description of a very satisfactory hen coop is reasonable, and I will try to crowd it in here. A box 2 feet by 1½ feet and 15 inches deep is about the smallest practicable, but any dimensions larger are feasible. Remove cover of box carefully. Select the soundest end and nail vertically two 1x2 inch strips 10 inches apart and extending 6 inches above end. Now nail a board six inches wide across the end by using those two strips. This gives you one end six inches higher than the others or the sides. Now mark off a doorway one-half inch inside of strips and ten or twelve inches high, saw out this 9x12 block and you have a half-inch flange on each side. Cut a piece of board 10x13 and hinge it with cheap hinges or straps to one strip and put a wooden button on the other strip, the door fitting in between them. Cut three strips of lath 13 inches long and two 10 inches. Lay the two strips down 10 inches apart from side to side, with the third lath evenly between them, and nail the two short pieces across the ends of the others, giving you a gate 10x13 with one slat in the middle. Now this will set in behind the hinges when the door is open and the button will secure it just as it does the door. For the roof cut two boards 6 inches wide and 2 feet 6 inches long, hold them parallel on edge 18 inches apart and lay a light roof on them, letting it overhang the sides about 3 inches. Cover this with any waterproof paper and put on lath at each side and end to batten it down. Now this roof will set down over the box with the slant from that extra 6 inches of height at one end and the side boards 6 inches wide will cover the gaps on side of box. Near the bottom of these same six-inch side pieces drive a single good nail straight through into box end, one on each side. That gives you a pivot that is secure and saves hinges. The roof lifts from the front like any box lid swinging up on those nails. Bore a row of large holes, whitewash inside and out or paint and you have an excellent weatherproof coop for hen and chicks. It is sometimes an advantage to nail a couple of strips across bottom to hold it up a little from the earth.

The nests in my houses are placed under the dropping boards and are made like drawers to slide in and out on a shelf suspended between the legs of the dropping board. Each nest is 12x13x12. A strip should be provided to prevent the possibility of shoving the nest drawer in too far. A hinged flap hangs down from the front edge of the dropping board and forms a front to the nests which can be raised to remove nests. The hens enter from the back and find a darkened secluded nest in which to lay and the possibility of broken and eaten eggs is very small. Nests arranged in this manner are as easily cleaned as any and do not encumber the floor.

—Trap Nests.—

To convert these nests into trap nests is a very simple matter, but I do not know whether I can make it clear without a diagram. I take No. 28 sheet brass, which is springy, and cut some strips 8 inches long and one inch wide, bend them at one end into catches by first bending up two inches at a right angle and then turning down about 1¼ inches into a slide so that when door swings down it will strike that slanting surface and push catch down until the corner is passed, when it springs up and secures the door from opening. These are attached to the lower side of shelf that makes nest bottoms, with two screws pretty close to the end away from catch so as to leave plenty of spring, and they project past bottom about 1¼ inches. Next take some No. 6 or 8 iron wire and bend it into square topped staples with one leg three inches long, the top 18 inches wide and the other leg 7 inches long, and then turn up one inch to leave a blunt point. The three-inch leg has one inch bent at right angles and pointing to the other leg. These wires should be hung with a small staple at each end and one in the middle directly on a line with center of each nest with the short leg hanging down about 12½ inches from nest fronts and long leg hanging above center of nests. The spring catches are under center of nests at front edge. Now hinge to platform six small doors 1 foot square, so that when they drop down they will latch on springs, whittle some triangular pegs and drive into middle of lower edge on door with flat side of peg down. Now when you swing this door up to set it the short edge of peg will push the short wire leg to one side by striking against the bent inch long end, and that end will slip under peg and engage the flat side, so supporting the door. A hen comes in to add to your bank account and jumps up into the nest of her choice; as she steps into it her head or body strikes the long leg of wire which swings to one side out of the way and in so doing swings the short leg from under peg and the door drops and latches. When you make your rounds, you lift the hinged board, lift out the hen and take her number, remove and mark egg, and pull down spring catch, then push door which swings up and sets itself for the next comer.

It should not be necessary to say anything about the desirability of trap nests for all poultry keepers, and for market poultry keepers they are frequently the means of changing a losing business to a paying one. It may not be possible to give the attention trap nests require to all your yards, but at least have the houses equipped with traps that are occupied by your breeding stock, otherwise you are more likely to run down your average egg production instead of increasing it. At the season when you are growing young stock to replace breeders, the heaviest layers are the ones that have laid fewest eggs during the rest of the year.

About the first thing your trap nest records show you is that the steady fall and winter layers are doing almost nothing at the season of natural production, so that when no records are made you are inevitably using a majority of eggs from your poorest layers and it takes a very short time to cut down your average egg production badly.

To return to the house equipment. This arrangement of nests enables one to clean out quickly and thoroughly by simply lifting flap and drawing out bank of nests, which allows nest material to fall to the floor. Under the shelter of platform at one end I put a shelf to hold fountain above floor litter, and at the other a set of grit and shell boxes; a feed trough for soft food completes the arrangement. The feed trough can be hung on pivots beneath nest boxes, but where many flocks are cared for the delay in feeding offsets the advantage. If the house is not warmly built, a curtain hung above front edge of dropping platform and falling to it will keep the chickens warm in severe weather.

I have gone thus extensively into detail, because as a rule the dimensions of fittings in poultry houses are ignored in the description of model houses, and that is the point where most beginners who are doing their own building make mistakes, and very likely I have taken too much space in discussing all equipments. Of course the veteran poultry man will adopt things to suit his own ideas but the beginner is so dependent for success upon proper equipment, and its lack has cut short so many enthusiastic poultry careers, that I may be pardoned for emphasizing the necessity of starting right.

In taking up the question of breeds for a market poultry farm, so much depends on the "man behind the hen" that I do not think any one person's experience can be conclusive evidence. But there are certain requirements in the business that narrow the opportunity of selection considerably.

Perhaps I should explain here that my choice for a title for these talks was rather haphazard, but was intended to include the production of eggs for market, as well as for broilers, frys, soft roasters, capons, old and young ducks, and the less attractive "hen for soup." However, in discussing the question of breeds I am forced to consider it from a fancier's standpoint, as well, for the sale of breeders, eggs and exhibition stock contributes no small share to the poultryman's success. Nine hundred and ninety-nine of every thousand fanciers depend for their daily bread upon some other avocation, while the market poultryman who succeeds must give his entire energies to the business. Therefore he must earn enough from poultry to support him. So it follows that he must raise large numbers of chickens, etc. Now, it is just as easy for him to raise pure-blooded stock, they are much more profitable per capita. And the question of fancy stock refuses to be left out, for it is inevitable that when a man raises pure-bloods of the utility breeds his neighbors will want to follow suit, and he becomes the source of supply for an ever widening circle of demand. Now, as it is essential that he will raise large numbers, it follows perforce that among the number there must be some better than the rest, and assuming his beginning was made with birds already on a high plane of excellence, the best birds he raises may very easily be super-excellent. When he reaches a point where this excellence is recognized in competition a demand is created for the best he can produce at profitable prices, and he has the great advantage over almost the entire brotherhood of fanciers, of large number from which to select.

—Selecting a Breed.—

We will now pass on to a consideration of the breeds adapted to a market poultry farm in all its branches. First and most important, they must be capable of laying winter and summer. Then they must give a good plump carcass at all ages from one pound through to old age, and without requiring a lengthy period to reach maturity, but must be of large size when matured. If all this points unmistakably to the American breeds it is because to these essentials the development of such breeds are due. In my opinion the Plymouth Rock holds first place by a comfortable margin and I prefer the Barred because they have been bred for so many years that some strains are unapproachable as all the year layers. They have, however, the disadvantage of dark pin feathers that are troublesome in the broiler stage, and there is no question that while the rewards are far greater to the breeder who approaches perfection closely, the difficulty of producing first-class specimens is equally great.

So that looking at the question from a fancier's viewpoint the beginner has much to accomplish before gaining recognition and it might be wise to take up the solid color branch of the breed. The nearest competitor of the Plymouth Rock is the Wyandotte, and to my mind their relegation to second place is due to smaller size and smaller egg

HOW TO MAKE POULTRY PAY.

capacity. They are as good as the best as broilers and frys, but as soft roasters and capons they lack a pound or more that might better be present. Even the Plymouth Rock can be improved upon as a capon by a Brahma cross, but while I practice this cross and caponize extensively, I am doubtful as to whether it pays to carry these capons to their greatest possible weight where all the food has to be purchased. That it pays to caponize simply to save food, I have no doubt, provided you can be sure that no birds are included that are worthy of being saved for breeders. The best time to caponize is at an age when it is in many cases impossible to predict a bird's quality and I have frequently sacrificed valuable birds in that way, so I am chary in advising it.

Thoroughbred Brahmas are entirely out of the question for a market plant in spite of the statement I noticed one month by Mr. I. K. Felch in Commercial Poultry of a hen of his which laid 313 eggs in 333 days. I have the highest regard for Mr. Felch and I am quite certain that he believes that to be a correct record, but I do not believe it and I doubt if there is one single breeder in the United States of any experience who would believe it. Brahmas would make fair broilers if it were not for their legs, but they are so long that it gives a spindly impression to the carcass. I suppose in writing the above I am opening the "vials of wrath" upon my head, but I am merely giving the "faith that is in me," and I must abide the consequences. After all, I have no special axe to grind, and while it is true that I have been known to sell breeding stock if sufficienlty urged there are lots of others who sell stock also.

To return to capons, if you have a private trade that pays you sixteen to eighteen cents a pound for large soft roasters, you cannot charge enough more for capons to make it an object to caponize and feed for the additional time, at least that has been my experience, although if your product goes into the open market I should judge the results would be the reverse. If, however, you have a larger supply of cockerels at two and a half to three pound weights that are palpably unfit for breeders than your broiler orders will use, it most certainly pays from an economical food standpoint to caponize and sell them as soft roasters at six to eight pounds, as nearly one-third less food will be required to develop them to that point.

When you consider that about every old woman in France who keeps chickens will successfully caponize with no other instruments than a pair of scissors and a hair pin, the simplicity of the operation is indicated. I have at times caponized entire flocks with no loss and at other times might lose 5 per cent., but there is no absolute loss if you select market day and then if you kill any birds under the knife dress them at once to fill the orders and if you stop when orders are filled and postpone further operations until another order day it holds you safe from loss. I speak of order days, because as a rule on my farm the dressing is done three days a week for convenience sake.

There is a time honored belief that for egg production alone the Mediterranean breeds excel and where they are allowed to roam at will with little care it may be the fact, but where yarding is necessary and practical feeding followed by experience doesn't bear it out. I have White and Barred Plymouth Rocks and Single Comb White Leghorns in yards side by side and during cold weather the Leghorns are way below the average, but during the hot weather they do a little better. The only drawback I have found with the Plymouth Rocks is just at the end of the moult. While moulting is in progress they lay very fairly, but when the new feathers are well out they are prone to fatten suddenly as a result of the rich food needed to keep them laying while moulting and that fattening shuts off the egg production at once. It grows easier each year to control and hasten moulting and at this date (October 1) more than one-third of my flocks have entirely finished, another third are well along toward completion, and the rest show ragged and dull with

VIEW OF POULTRY FARM OF COLUMBIA SCHOOL OF POULTRY CULTURE, WATERVILLE, N. Y.

more or less dropped feathers. I have been impressed this year as never before with the fact that a hen that is set in July will almost always moult in her coop and rapidly grow a new coat so that when chicks are weaned she is already laying in her coop.

Like lots of other poultrymen I have gone through a succession of bone cutters in a vain attempt to find one where no work is required to cut an adequate supply of green bone. It is all very well to feed bone where a few dozen hens and chicks are kept, but unless a source of power is available it is out of the question to cut green bone in sufficient quantities for hundreds of hens and thousands of chicks. That has forced me to using one of the commercial beef meals and I find it absolutely indispensable. With the high grain prices this year it has cost me pound for pound less than oats or corn, and about the same as wheat, while its food value is enormously greater. I use several hundred pounds each month and I shall show the biggest cockerels this winter that I have ever grown in consequence.

Beef meal and boiled oats give tremendous bone and muscle. A word of caution in its use will not be out of place. In beginning to use any new food it should be fed sparingly until the birds are accustomed to it and then gradually increased to almost any quantities, watching to see that they are assimilating it properly. As a rule the guaranteed analysis falls far short of real protein contents with these foods. Another result I charge to its extended use on the farm this year is the continued laying of my Pekin Ducks right through the prime moult and far past their usual period. I have two ducks that are still laying daily, and those eggs incubated will give me ducklings when they bring fifty cents a pound.

—Keep Some Ducks.—

As a matter of fact no market poultry plant is making the most of its opportunities unless ducks are kept, for the yield the surest revenue of any fowl I know, for the least outlay of care and attention. The cheapest sort of a shed is sufficient for the breeders. The eggs hatch well and the ducklings require a far shorter time in heated brooders and are ready for market sooner than chicks. They are not subject to the host of ills which frequently cause heavy losses with chickens and their food required is the plainest and cheapest. Bran, cornmeal, beef scrap and a little forage makes their entire bill of fare from start to finish and the feathers more than pay for dressing them. At least, the Pekins do, for white feathers bring me fifty cents a pound and four ducks produce a little over a pound. Only water for drinking should be allowed them, and that in a vessel deep enough so they can clear their nostrils, otherwise they die. If the breeders have a pond, stream or slough they will feed themselves most of the year.

Moulting.

There would be nothing unusual in having the molting nearly finished by October 15th. By that date 5 per cent. of my flocks had not completed their molt. In this connection I wish to speak of a, to me, very remarkable occurrence. On October 11 I was showing some White Plymouth Rock hens to a prospective purchaser, and one in particular attracted his attention. She was a ten-pound hen with two years' experience of show coops and handling and very tame in consequence, so she did not struggle or become excited in the least. Apparently she had not started to molt, was not ragged or badly weather-worn and

A Low Priced House With Covered Runway.

feathers did not come away in handling her. The purchaser planned to come two days later and get her and another hen selected. The following morning when I entered that coop my first thought was that a cat or skunk had been there before me, for at the end of a roost which was this hen's particular haunt, was a mass of white feathers, as if a pillow had been emptied there. There was no carcass, however, and I passed hastily on into the scratching shed adjoining to see that hen energetically breakfasting with her entire body as bare as the back of your hand. She looked enough like a dressed capon (or undressed) as any ready for market. A ruff of short feathers around her neck, most of the flight feathers and stiff tail feathers, and dainty little garters of white, completed her morning costume. She seemed rather proud of it, but I gathered her in and put her in a training coop in a warm room with a chunk of raw beef to entertain her.

By night she was covered with a greasy, milky growth of pin feathers and seemed a trifle distressed. In the morning her covering was a sort of quill armor with here and there a tuft of feathers sprouting from a quill point. She seemed very restless and I put a large pasteboard disk about her neck to stop her working at her quills and floured her heavily, to soothe the irritation. She fed voraciously, but was evidently losing weight fast. I provided all she could be coaxed to eat of the most concentrated foods, raw beef and beef suet and oatmeal cooked with cream and whites of eggs added to make a stiff paste, and rolled into pellets she could swallow easily. I considered it a toss up whether she would die of exhaustion from such an abnormal molt or indigestion from such rich food. I had placed her on the scales for my customer, and I weighed her again at noon, twenty-four hours later, when she had lost one and one-quarter pounds. All day her feathers sprouted almost visibly from the quills and at 10 o'clock that night she was gasping with nearly two degrees of fever, and greatly exhausted, but considerably fledged all over, with most of old wing quills dropped out, also tail. I gave her one grain each of Codine and Mur. Ammonia, put her in a stout paper sack with holes for her head and feet, and went to bed. In the morning she was too weak to stand much, but still ate freely. Her buyer arrived, and I gave him another in her place, as I wished to see the finish. About 5 o'clock she refused to eat, but temperature was nearly normal, and after providing for her warmth, I left her until the next day, when she was stronger, ate freely, and had some new feathers two inches long from tip to skin. She made steady growth of feathers and strength, but had now lost two and one-quarter pounds. Nine days from the beginning of this curious molt she was returned to the yard with a fairly complete, though, of course, unripe, coat of feathers. Ten days later she had regained three-fourths pound of lost weight, and was thinking of laying.

I would give a good deal to know the cause of such an abnormal molt, though she would undoubtedly have died without close attention, yet if such a process could be induced it would be of great value. So far this year I have not lost a bird, and I attribute it to two reasons. First, to the very narrow ration that I feed at this season; second, to the fact that in former years in following out my ideas of maintaining the highest possible average of vigor in my flocks, I promptly retired from business any hens that showed weakness and debility during the molt. I do not insist that a hen shall lay during the molt, though I naturally prefer it; but if she does not, she cannot take many rests during the remainder of the year without falling below the annual average I require. If she does fail she retires to grace the stewpan, and so fulfill her ultimate destiny.

—About Some Houses.—

I have one that is an example, for I gave a good deal of thought to it before I built it, and it was an utter failure as a brooder house, for which it was designed; still it is by far the most expensive building I ever put up for chickens. I will give the dimensions and explain wherein it fails as an object lesson in what not to do. The main house is 16x32 on the ground, with 16x10 annex as an incubator room, which last is all that can be desired. The main house is 3 feet, 6 inches at walls and 6 feet 6 inches at ridge inside, is double sheathed, building paper on both sides of studs and ceiled inside, all of matched stuff. The sash are 2 feet 6 inches square and are double glazed with one-half inch between glasses. Three-ply building paper on roof, and the door is made refrigerator style with three thicknesses. Foundations are broken brick, laid up with tamped cinders. Floor is of alternate layers of cinders and clay tamped hard. Six windows in each side give plenty of light. The house stands north and south, and I figured that the low windows would bring lots of sunshine in on the floor morning or afternoon, but I was badly mistaken. The comparatively low winter sun gives its best efforts from about 10 a. m. to 2 p. m., and during that time is so much to the south that only narrow shafts of sunshine come in on the floor and it was entirely impossible to sun the hovers properly, as they were ranged along the passage through the middle of the house. Large skylights would correct that, but even with double glass the condensation is so great they constantly drip, which make them impracticable over chicks. As a house for stock it is not so bad, but would be improved by raising the brick foundation two feet or so, which I expect to do. It has too, the disadvantage of being too warm, unless you confine your birds through the winter, and I do not believe in that. If hens sleep in a very warm room and then they are let out on bare but frozen ground they are exceedingly likely to take cold from the sudden changes of temperature. With this the house cost nearly three times what a chicken house of equal capacity should cost. The incubator annex is built with an additional layer of paper, boards and air space, with a large air box opening at the floor on the north side and two flues

HOW TO MAKE POULTRY PAY.

against the roof on the south, which gives an incessant flow of pure air into the room, an essential to successful incubator operation.

To get back to market poultry raising, every farmer who raises poultry and sells it to the chicken buyer, the local market man or to private customers, effects the price of poultry and generally depresses it, because his birds are ill bred, ill fitted and ill dressed. Few farmers realize that the difference in weight between a flock fast reverting to the jungle fowl type and a flock of purebred or graded quality would in dollars and cents not only purchase purebred cockerels, but in most cases pay for a pen of good quality and leave a balance besides, while the second year's crop would represent in increased weight just so much absolute profit, for the difference in food consumed between free range scrubs and free range fullbreds is too small to consider.

Most farmers are understocked as to range and fearfully overstocked as to houses. That question of breeding up the quality of farm flocks is one of the first forced upon the attention of the market poultryman, because the farmers about come to his place and try to trade scrubs for fullblood cockerels. The farmer sees him marketing stock about double the weight and three times the value of his, for every housekeeper recognizes that there is less proportionate waste in one big than in two small chickens, and have surplus cockerels to sell as broilers, roasters and breeders, and so provide funds to carry on the work. With the advent of February I would bend every energy to the possible number of chicks from this really good parentage and continue hatching as long as the flock provided eggs and I could pick up sitters cheaply and either reset them or start them laying and make their eggs pay their cost. When my year was up I should have a considerable number of extra fine cockerels and my original pen. Out of all these, reserve the very best and sell enough of the remainder to pay for housing and winter food for the flock. By that time my early pullets would be laying and a few customers for dated breakfast eggs, at say forty cents a dozen, would give a revenue, the size of which depends only on my success in raising plenty of early pullets to lay all winter. In the spring I can advertise and sell sittings and possibly a few breeders and devote myself to producing a second and much larger crop of chicks and with good management the business is successfully launched on a paying basis.

For the first few years my advertising should be as extended as possible and regarded as developing the business rather than as calling for immediate and equivalent return. It takes time, time and reiteration, to present yourself to the purchasing public, and you must remember that the field is already exploited and only the best methods can re-

Light Brahma Chick, Bred by Frank P. Johnson, Sta. A, Indianapolis, Ind.

where he comes into competition the farmer's stock has no chance. Such an object lesson can not be disregarded, so he tries to trade with you or purchase elsewhere, and so the leaven of improvement is introduced. That demand for improvement, strictly local perhaps at first, spreads through a wider and wider circle of influence until the market poultryman becomes the nucleus of a district where superior poultry is produced and he is perforce driven into the ranks of the fancier because his own birds must constantly be improving to supply this demand for better birds.

The earlier in his career the market poultryman realizes this, the easier his road to success. It is not a difficult matter to make poultry pay, and pay well, but it is a mighty tough proposition to have to live on what your poultry earns the first year, even if you have lots of experience, and without experience failure on that basis is certain.

If I should start a market poultry plant to-morrow with $100 capital and food and shelter provided for myself for a year, I would divide my capital about as follows, because experience has shown it the easiest and most direct way to success: Twenty dollars for a warm little house and roomy scratching shed, $60 for a cock and five pullets or hens of the utility breed I find most attractive, and $20 held in reserve to feed my flock and their progeny, purchase sitters and incidentals until such time next fall as I

main at the top. Aside from any question of personal honesty, if I write an intending purchaser that I will sell him better stock than another can afford to at the same price, and he does not believe me, I lose the sale. If he believes me and I fail to make good my business reputation has suffered vastly more than the amount involved, but if I do send him birds of the quality promised and they are better than my competitor's at the same price, I have made a friend that will boost my business methods when he is talking chicken to his friends, and that is the healthiest and most satisfactory growth of business reputation and worth many times over an advertisement declaring I breed the best on earth, etc.

There is a question that forces itself upon the attention of the market poultryman who raises pure-bred stock. After you have created a demand for fine table poultry, the greatest struggle is to keep pace with the growth of that demand. On paper it is easy to demonstrate that by setting so many eggs in incubators, hatching such a percentage, and raising to marketable size another percentage of those hatched, you will have an unfailing supply at regular intervals. When you try to reduce these calculations to practice you find that no arbitrary percentage will cover the fluctuations that occur in actual work, and you are over-producing or under-producing, perhaps not in the year's total output, but in the immediate supply to satisfy your orders.

HOW TO MAKE POULTRY PAY.

For the sake of demonstration, we will say that something has temporarily gone wrong with your breeding stock, but not sufficiently wrong to be immediately noticeable. As usual these eggs are incubated, but fall far below the percentage expected to hatch and those that do hatch are pretty sure to show a correspondingly decreased vitality, and so the greater mortality has perhaps cut in two your expected supply at a certain date and the hatches following follow the same trend until the correction in the condition of breeders has taken place. If your orders have been absorbing your entire output, there is a sudden gap in the supply. What will you do? An essential to the building up of a private trade is absolute regularity. If you fail your customer, she seeks another supply and you have lost valuable business. Apparently the remedy is consistent over-production so as to establish a safe margin, but if your degree of overproduction is too great, the additional food expense swallows a large portion of profits. If in place of overproduction you attempt to fill gaps by purchasing the birds for marketing, there is prompt trouble, because of the lowering in quality. And you are only enabled to charge extra prices because of extra quality.

Now you are raising pure-bred stock of a breed that makes a gilt-edged table fowl, and if you can fill gaps by purchasing birds of the same variety, your task is much simplified. To this end you will find it pays to select from among the farms about you, those where the equipment and surroundings are best, and stock them with pure-bred stock, under an agreement with the farmer that you shall have the refusal at the current prices of all poultry and eggs produced that you desire to take. In practice, if you maintain friendly relations, such an agreement works very well and you are protected from competition by your own stock. It also has the advantage of giving a place where stock can be pastured or given a vacation when such stock is of too high quality for marketing. When for any reason you have a gap to fill there are always these birds of your own breed to fall back upon and bridge it over until your own supply is adequate. There will be many cases where people you have not intrusted stock to will come and buy in order not to be outdone by their neighbors.

In only one case in my experience was any attempt made to defraud me by selling settings without my knowledge, and in self-defense I was compelled to resume the flock when, after a short interval, that farmer purchased sittings and launched himself as a breeder of pure-bred poultry, simply because an experience of my flock showed him in dollars and cents that he was not making the most of his opportunities. Dollars and cents talk in a very earnest way and their arguments seldom fail to convince. The only trouble is to bring the argument home to the farmer. That requires practical demonstration; he does not believe all he is told or reads (not even about gold bricks), but when he has been interested in the possibilities and can see for himself what the difference is between the profits of pure-bred and scrub, then he becomes a customer.

For a case in point: Four years ago a farmer drove into my place from his farm a good many miles away and tried to buy some cockerels at fifty cents each to grade up his scrub stock, which were all dunghill runners of the most reverted type. I had none to sell at any such price, but I took him to a farm a short distance away where I had given them two cockerels in return for many kindnesses, and there he saw a flock, originally of the three to four pound, black and brown scrub type, rapidly increasing in size and assuming a distinct character as grade Barred Rocks, the result of once breeding back the original cockerels to their pullet progeny. He did not say much, but the lesson was clinched when the farmer's wife made a trip with her egg basket and came back with it full, and that at a season when most flocks were not laying at all. He came back to my place with me and rather grudgingly gave up $5.00 for three cockerels. That was in November, and the next year, in September, I had a note from him asking if I would buy his chickens. As it happened I had a butcher who was begging for stock, and I wrote the farmer that I would take all his pullets and cockerels. He appeared a few days later with several coops of very decent marketing chickens and after weighing them in, I was about to pay for them when he said he guessed he wanted to take it out in stock, as he had decided to raise nothing but pure-bred birds, and he did. He drove away with the nicest flock I had sold up to that time, after looking at different styles of poultry houses, as he intended to build a new chicken house. He has been a steady customer ever since and has kept a very accurate record of receipts and expnditures of his flock, with the result that his poultry has proved nearly twice as profitable, for the amount invested, as his herd of dairy cows.

He has shipped five cans of milk daily for a good many years, and the labor of himself and two grown sons, and almost the entire product of his one hundred and sixty acres is required to provide that milk supply, which averages a net income of sixty cents for each eight gallon can, or about $1,100 annually.

That certainly seems to me a "strenuous" way of making a living, but doubtless more skillful management of his dairy herd would yield larger returns. His herd eat up everything grown on the place, aside from kitchen garden and a small potato crop. I will venture to say that if I were trained to general farming as he is, I could turn the products in grain and forage of one hundred and sixty acres into poultry products at a profit several times greater than his entire income.

Enthusiasm is a great factor for success in almost every vocation, but it is a curious fact that the "hen fever" which produces the most rampant enthusiasts, is much more serious when complicated with enthusiasm. If I were writing a "hen dictionary," I should define enthusiasm as the acceptance of vague possibilities as facts, and the utter disregard of probabilities.

If there was no money to be made in the poultry business, a good many of us would try something else, but because a good many do turn from the business to something else, is not evidence that money cannot be made with poultry, but merely that the party retiring has been unwilling to accept probabilities, has ignored details, and, losing interest, has lost capital. These doleful remarks are prompted by the fact that I was recently called in to attend the obsequies, in a professional sort of lay out the corpse way, of a poultry plant that I referred to last spring as a shining example of why there were failures in the poultry business.

As usual it was the failure of the man, not the business, and it points the moral I wish to make because it is such an ordinary, unnecessary, useless cause for failure. Merely the lack of any system of accounts or of any means of comparison between receipts and expenditures. He declared he had a system, and he had, if you call it that, because his bank balance showed that at the start he had a certain sum, and now he had not. I don't call simple subtraction a system.

Of course there are people in all sorts of business who never know where they stand, and it sometimes happens that their ignorance continues until they precede the sheriff through the door and watch him lock up, but more and more each year a fairly close system of accounts is put into operation, even in the farmer's business, the agricultural school graduate is introducing system. And a certain degree of system is essential to any considerable success with poultry, not only system in cleaning up, feeding, etc., but system in handling your letters and systematic accounts.

To the market poultryman above all (in distinction from the Simon pure fancier) are accounts essential, because from year to year, and season to season, the cost of foods vary widely, much more widely than the prices received for his products, and in these variations lie his margin of profit. It is only by comparison that tangible results can be obtained. Personally my records are very simple, for I have not the time to give to any elaborate book keeping.

An incubator book that shows date machine was started, number of eggs set, number tested out, number hatched, and the number of brooders to which chicks were transferred, together with the daily thermometer records. A brooder book, that under numbers of brooders, shows when chicks were hatched, what feeds were used, what mortality, and, as far as possible, why, general notes as to growth and condition, and finally what disposition is made of the chicks, whether transferred to colony houses or put into broiler pens for marketing.

—Keeping Records.—

A stock book that shows under numbers of pens, the number and variety, the numbers in colony houses and the egg records. For this last a page is given to each hen, showing her band and pen number, the page is ruled vertically into twenty-four spaces or two years, and horizontally into thirty-one spaces for the days of the month.

HOW TO MAKE POULTRY PAY.

The date hen was hatched is at top with band number; when she begins to lay the name of the month is written in first space, and on the numbered line corresponding to the day of the month a figure one stands (mine never lay more than once a day, I am sorry to say) and from there on in regular order as the eggs are laid; at the bottom of column a space for totals show each month's work. Pages ruled from Humphey & Son's catalogue are used for monthly records. This may sound rather formidable, but as a matter of fact aside from the trap nest records, a pocket memorandum book will hold it all. A ledger and a day book that show daily expenditures, as well as daily receipts, are also needed in the business. Then to handle the correspondence, an ordinary commercial copy book, letter files, and a card index are a great help, and you don't have to depend on memory. And while I am speaking of system, let me say this, never write your reply across the foot of a man's inquiry and return his letter with it, for you may want to recall what he asked and you certainly should have a copy of what you answered, besides that letter may cause embarrassment some day if you see it published in the poultry press with caustic comments by the editor, as happened to one unfortunate a few weeks since, and in any case it is a slovenly way of doing business.

If you are hatching and marketing chickens the year around it is rather hard to strike more than an approximate balance sheet, because the food consumption, aside from the laying stock, does not only represent the chickens marketed that same month, but also the chickens preparing for the next two months or so.

After a season's records are at hand that show the cost per pound, let us say of chickens ready for market, then an approximation can be fairly accurate; thus December, 329 chickens, at a cost of 24 cents each to produce, $78.96; 143 dozen eggs, at a cost of 7½ cents to produce, $10.73;—$89.69; sold 329 chickens (broilers) at 60 cents, $197.40; and sold 143 dozen eggs at 40 cents, $57.20; total, $254.60, less $89.69, total expense, gives $164.91, net profit.

Now you know you made a profit last month even if you did spend $212.00 for grain and mill feeds. And it is very comforting to know you have made a profit in spite of your expenses having much exceeded your receipts. It gives the "eye of faith" a chance to see that when that $212.00 worth of food is consumed, your February account will show about 1,100 broilers to be sold at 75 cents each, etc.

Such a system of accounts and records gives you a sure way to stop the leaks and swell the profits. If any particular season shows business done at a loss, analyze it, find out why; if you cannot make a profit at that season stop trying and rush things harder when you can.

Here is my confession: I don't hatch any more in the last five or six weeks of each year. Others may make a profit then; I don't seem able to make it worth while, considering mortality and vitality, value of eggs and cost of fuel, beside the other work at the rushed season. I thought for a long time that it paid, but my records showed it did not, and I quit trying. Find out for yourself.

First, in importance, is the renewing of your supply of early fall layers, and that means pullets hatched now and given every advantage for growth and comfort. It really begins back of the hatching, for you should not only select the eggs of your most prolific layers, as shown by their pullet records, but the care they are given has everything to do with your ultimate results. If those record laying hens are too fat now, they will be laying just the same, but their eggs will give very poor results, not only in the incubation, but in the ability to thrive (and that means something more than merely keeping alive), after being hatched.

This is one of the points about the poultry business where experience is essential. The best of poultry literature can do no more than warn and advise or point out the necessary conditions for success; with the poultryman himself the whole matter lies, and he must learn to recognize the proper condition of layers to insure the highest fertility, and frankly it is not very easy to learn that. There are lots of people in the poultry business to-day who are utterly at sea when it comes to knowing the proper condition of layers whose eggs are to be incubated, and this accounts in part for the very unsatisfactory results that sometimes follow the purchase of sittings. For a case in point, no longer ago than last spring, I purchased from a fellow breeder 150 of his best eggs at a very decent price. I knew the man and knew his birds but I unfortunately did not know how he cared for them. The eggs arrived and after testing them I put them in trustworthy incubators, with the exception of three settings under hens for convenience in pedigreeing, as the machines had no pedigree trays.

I am afraid to say how many thousands of eggs I have tested in incubators, but my experience has trained me to recognize clearly certain conditions, and in this particular case as soon as I had finished testing I sat down and wrote my friend that his stock was so fat that the eggs were practically worthless, for while the test showed nearly 74 per cent. fertile, it showed just 11 per cent. of strong germs, and in reply he wrote that he thought his birds were too fat and so he sent me two extra sittings to make it right. Well, ——, I am not a kicker and I do not expect too much in buying eggs, especially incubator eggs, as this is only a lottery in any case, and so I held my peace and in due time hatched thirteen chicks from 170 eggs. They were weak, but I was so determined to make them live that I did succeed in raising five, but they were excellent copies in miniature of the breed they represented. I looked forward with interest to seeing my friend at the Chicago show, and I did, and because I did not tear him to pieces he grew confidential and said that he had never had so poor a season, little would hatch, and what did die "so easy." And he has been in the poultry business for years, yet has failed heretofore to realize the essentials for strong fertile eggs, and so lost an entire year out of his business. This is one reason why so many breeders at the close of each season declare they will never sell sittings again, for it is pretty bad on a man who is perfectly honest in intention to be scolded and called names simply because of his ignorance. That is why I say that in this particular, experience is essential as well as in some others. There is much that can be learned through the poultry press, but some things require the absolute experience and demonstration.

"To get back to our knitting," if your layers are in proper condition, you have come a long way toward success, but the neglect of other things will still prevent results. There is as much difference in the advice incubator manufacturers give as there is in their machines, but you can take for absolute fact that thorough airing and cooling is essential to any machine. If you happen to set a hen who will scarcely leave the nest for a moment, you can be pretty certain she will give a poor hatch as much as the hen who goes off for hours at a time, though if the weather is not severe, the latter probably will hatch more and stronger chicks. There is one mistake that many people make who should know better, and that is putting the weakly or crippled chicks into the brooder with the rest because of a wish to save as many as possible; that is the very worst way to go about it. Either kill the weakly and crippled outright (by far the most humane way) or put them in a separate brooder so they do not jeopardize the health of the remainder, which they are sure to do because the mere fact of being weakly predisposes them to disease and the first exposure sickens them and helps to start sickness among the others, besides the fact if they avoid death they are never worth anything, and if they die at night in a brooder, the high temperature very quickly makes them decay and poisons the air which their fellows must breathe and so weakens them and prepares the way for more disease and death, till what survive are not worth saving.

S. C. RHODE ISLAND REDS.

First prize pen at Illinois State and Indianapolis Shows. Bred, owned and exhibited by Messrs. Murray & Wheeler, of Springfield, Illinois.

WINTER EGGS AND HOW TO GET THEM

An Effort to Interest Farmers—Rearing Pullets to Become Winter Layers—There Is More Difference Between Different Strains of the Same Breed Than There Is Between Different Breeds—Feeding, Housing and Caring for the Layers.

By HARRY J. WOLSIEFFER, Egg Harbor City, N. J.

MUCH has been written on this subject and it cannot seem a new one to old readers, still there is courage installed in the beginner and new confidence put in the older poultry raisers when they read of the experience of others, who are having more and more success each year. There is no question but that there is yet much to learn in this getting of winter eggs, not merely from February on but from October or November until summer comes along. It is not the beginner alone who experiences difficulty in getting the hens to lay when eggs are high. Greater egg yields and longer winter laying are becoming more common every year and our journals do well to give attention to this branch of commercial poultry, although I see that some of them still neglect this branch for the fancy. It is the writer's opinion that the more these practical subjects are talked over in the press the more confidence is given to the beginner. The farmers of this land should be considered first in the poultry industry. As a rule they are hard to reach, but once reached and convinced of the possibilities of the business they themselves make it many new converts. Poultry raising with the farmer is generally a side line. I know of no one better fixed to make money out of poultry than the farmer with every condition favoring him, and it has been the hope of the writer who is a farmer (but leaning more and more every year to poultry) that the poultry journals would find their way more each year into the homes of farmers. In time our own farm will, we hope, contain nothing but fowls and fruit trees. We started farming, but found that poultry paid better each year than any farm crop; a sure crop, one that could be counted on—in spite of the warning, "Don't count your chickens before they are hatched."

"How do you obtain winter eggs, and can others do it?" is a question that comes to us often. It is an easy yet hard question to answer. We started with six pullets and a cock some years ago. That was the foundation of our present flock. We paid a good price—not one bit too much when it was the foundation of our success or failure. Those birds were carefully raised by an experienced breeder from a heavy laying strain. We paid for that breeder's experience and years spent in perfecting that strain; we started where he left off, and with stock just as good. And in this experience is our first advice to the beginner, and the one least often carried out. Obtain the very best, either for utility (eggs and meat) or for the fancy. Be willing to pay for the best. It is a saving in the end and prevents the loss of several years' time, for should poor stock be obtained and the breeder stick to the business they will in the end be discarded for good stock.

We had no trap nests that first year, but every egg was carefully gathered and at the end of the three hundred and sixty-five days we found that the fowls had proven unusually good layers. The average satisfied us and we bred from all these pullets and have never regretted the transaction which brought them to our hands.

All fowls if carefully fed and cared for can be made profitable, giving a fair return for the money invested. But some fowls and especially some strains can be made to almost double the earnings of haphazard-bred stock.

—Rearing Chicks to Become Good Winter Layers.—

April and first of May hatched chicks should be good winter layers if grown properly without the usual check. Every time that the chick receives a set back it requires a longer time to bring it to maturity. Even experienced poultrymen sometimes overlook this fact and cannot understand why their pullets are not beginning to lay.

Proper housing, proper feeding and the proper strain of fowls are three essential points in getting winter eggs. Don't get off the handle on wide-open houses; go slow. If tried in some climates one is apt to get no eggs at all. We believe in fresh air; we believe that there is something in the open-house theory and each year we have tried it. But here in New Jersey, with much rain, sudden changes from warm to cold and from cold to warm, there is yet something to be learned and we who have during the last few years been obtaining eggs when eggs were high have not done it by giving them constantly open houses, but by using our judgment of when to and when not to open wide.

Where curtained roosts are used we would suggest that the fronts be made of muslin and glass, for the chickens like sunshine and there are many days when they can enjoy it only through glass. We use two windows, one of glass and when the weather is not suitable to its use it is slid out, and the other, which is of muslin is inserted in its place. This system gives us sunlight, fresh air and winter eggs. The poor results obtained by many are caused by improper breeding, improper housing, improper care and improper stock. By improper stock we mean fowls that have no records for good performances as layers. We venture to say that over ninety per cent. of the farmers and many professional poultrymen do not know how many eggs their hens are laying. They do not know which fowls are laying sixty eggs per annum, or which are laying two hundred. We have found that the best paying results are obtained only by breeding from those females that have proven themselves to be good winter layers. We need not add that the good winter layer, the one that lays many eggs during the whole year is a strong, healthy bird profitable to breed from. It is evident, then, that the egg farmer should use only those fowls that are bred from known egg-producing ancestry. Breed with egg-production in view and in a few years a strain can be developed that will considerably excel the average run of fowls.

With the birds five to six and one-half months of age the first eggs may be expected provided that the birds are well matured. We have had Wyandottes to commence laying at five months and we would not breed from any fowl that did not lay before the sixth month was up. Each year carefully select the best layers as near the Standard requirements as possible, but without sacrificing utility for beauty; for if the aim is egg-production utility comes first, beauty afterwards. If the two can be combined it is a very fortunate matter. The beginner should decide to give the larger part of his energies one or the other and should not hope to excel both a fancier and practical poultryman with the same strain of fowls; one should be a specialist either of one branch or the other, but specialism does not mean that the other branch should be ignored entirely. The fancier must pay some attention to the practical qualities of his flock and likewise the utility poultryman cannot afford to lose sight of Standard requirements, particularly those that relate to shape.

—The Value of Strain.—

There is as much difference between strains of the same breed as there is between different breeds. Because a hen is a Leghorn she is not necessarily a good layer. She must have been bred from an egg-producing strain before it can be prophesied that she will lay according to the reputation of her breed.

With us there has been no egg type, although we have noticed that many of the best layers were broad-backed, full-breasted, always active, of a nervous temperament and good eaters.

We have always used as breeders strong, active and perfectly healthy birds; we would not breed from a bird that had been sick even one day. A hen that is intended for a profitable layer must be in perfect health, have a strong constitution, as prolific laying is a heavy strain on the fowl and only vigor and vitality can pull them through the laying years in good shape, for the proper bred hen has not, as many writers claim, only one laying year but two and we have kept our best layers three years and have not lost money in so doing, they doing close to the 160 mark, which is a paying one. In ending the subject of strain we would say that heavy laying strains are not made in one year, but are built up after some years of hard work and constant care with much patient waiting. There are many poultry raisers we know of that are not known to the readers of the poultry journals, who are quietly working out the problem with trap nests. They have started with the best strain possible and are meeting with success. Each year new recruits are found in the ranks. Some fall by the wayside, but through no fault of biddy. There are several plants here around Hammondton and Egg Harbor that will be heard from some day, for it is the large plant that is most known to fame, although there are more of the 500-hen capacity that are making money in this State than of the larger ones. There are, however, many of the large plants which are making money. We had experience as manager of one that was making money and could have continued to do so had the owner left the management to a poultryman, but when several try to manage a poultry plant and one or more of the several have no practical experience, there is very apt to be another poultry failure charged to the already long list. That there is money in poultry in a large or small plant is no question with us. The success or failure is due simply to management—the better the management the more the profit. In the management lies it all.

—Feeding the Pullets That are to Become Winter Layers.—

For the first five months our fowls are fed for strong, healthy growth and range over fields of cow peas and under the shade of peach and apple trees. Oats have formed the bulk of their ration, with little or no corn, meat at the rate of ten or fifteen per cent.; barley and buckwheat have been fed for changes. We have taken no chances in the meat line—that on free range the fowls could pick up all they needed—we do not think so and our birds this year have made stronger growth than ever before. At five months they are put in the regular laying house, one which is 40x40 and eight feet high, with a capacity of 300 layers. This house was in former years a wine cellar, which accounts for some of the dimensions; but we have altered it so that now it is all that could be desired. It has two windows in each pen, one being of glass and the other of cloth, which is left in place the entire winter except during hard, driving rains, when a wooden slide is placed in front of it. The hay loft above gives good ventilation. The muslin window gives air and the glass sunshine. This house is lathed and plastered and the 160 feet cost us only $15.00. This gave us a double wall all around at a cost much less than to board and paper same. This plaster wall is a great advantage in many ways. Fighting lice is made easier, whitewashing is quickly and effectively done, the whole building has a cheerful appearance and the fowls have always shelled out the eggs to our satisfaction. On the south of this 40x40 house are four pens, each 10x25, in which fifty Wyandottes are wintered. On the north side are four smaller pens, a four-foot alley separating them from the others. We know that in placing fifty fowls in one room that we are exceeding the number that most poultrymen favor for best results, and to the beginner we will say that a smaller number will be better until the handling of fowls has been tried, but to the egg farmer who has many fowls and long houses we can say that with the same careful handling and room the results will justify having fifty in a flock. We have had the same hen average in flocks of fifty as in flocks of twenty-five. However, as a rule, the same careful attention is not given to the individual in the larger flocks that is given to her in the smaller ones. These large flocks we call our commercial pens. The eggs are shipped to New York and Atlantic City. No trap nests are used in these pens and they are forced for heavy egg-production during the two years that we keep them. It lessens the work in feeding, watering and in many other ways to put the fowls in one or two long house and in flocks of fifty. The yards are fifteen feet wide and 150 feet long, with peach and plum trees in the middle of each yard. This gives a natural shade and allows us to get in the yard two or three times a year with horses and plow. It does the fowls good and the trees also.

In the smaller houses and yards are kept the pullets that are to make their records for future breeders. They receive different care and treatment from those in the commercial pens and are put under trap nests to determine which is the 200-egg hen and which is the drone. The commercial fowls never get out of the laying houses during the two years that they are kept except for a period during moult. The breeders are treated differently. We give them all the liberty possible.

—Feeding the Layers.—

Our method of feeding is neither new or startling, being much like many that have been recommended in the poultry papers from time to time. We use less corn than many. In summer we use none, feeding green foods, oats, wheat, meat, millet, barley, bran and middlings. On this ration we had the best egg yield during June, July and August of our experience. Why should a heating food be fed in hot weather? We use corn in winter for a "heater" and not so much for its fattening value; the colder the night the more corn we feed. In normal winter weather we feed it always at night. It is fed cracked. In very cold weather whole corn is fed. Sometimes it is warmed in large baking pans. We do not take much pains to feed a scientifically balanced ration. Our balancing has been done according to the results obtained with any ration. Our present methods of feeding give results which justify us in continuing them, although they are deficient in several elements which are contained in the best recommended scientifically balanced rations. In order to make a ration that is properly balanced it is necessary to do a little guessing, for there is no rule by which one can determine absolutely in advance the nutritive ratio of a combination of several foods. The best ration is a problem for the feeder and he must keep changing about until the desired results are obtained and then keep everlastingly at it. A good ration which we are using is:

Clover and alfalfa, two parts; meat scraps, one part; bran, two parts; middlings, five parts; corn, six parts; oats, two parts. In this mixture the alfalfa scalded will make what is called a wet-dry mash, without the middlings becoming doughy. The best way to feed this ration would be to feed the alfalfa, bran, middlings and meat scraps in a mash at one time of the day and the wheat, oats and corn at another feed. Cabbage, beets, carrots and other green food should be fed frequently. If possible use alfalfa meal as the bulk of winter green food. It is much richer in protein than clover and results will be better by using same. The ration that we used last year with very good results was as follows (the amounts specified being for fifty Wyandottes): Morning, two to three and one-half quarts of wheat and millet, very little millet, just enough to keep them scratching in the litter for the small seed, of which the fowls are very fond. Noon—mash (five quarts), made of bran, middlings, alfalfa and meat scraps. Evening—cracked corn (when very cold whole corn) two and one-half to three and one-half quarts when weather is cold. When very cold four quarts. By feeling the fowls' crops at night one can determine whether or not he is feeding enough.. They should on very cold nights have very full crops, but when the weather is warmer not so much should be fed.

In the noon mash three quarts of alfalfa, three of bran, one and one-half of middlings and one quart of meat scraps is about the right proportion. At times this mash has been increased and at others decreased according to the amount of green food being fed in cooked or raw state. Here is where the judgment of the feeder comes into play. No two flocks will eat the same amount and good judgment with the feed pail must be exercised. This is especially true if the poultryman wishes to get down to the $1.00 per head cost of feeding. Feed enough, but not too much, and if the proper discretion is used even when buying all food one can feed for about $1.00 per head. For the past few years this has been our food cost per hen except last year, when feed here was the highest since we have been in the business, it costing us $1.10 per fowl and an extra hard winter helped to run the bill up, for heavier feeding is the rule in colder weather. In grains we feed wheat, oats, buckwheat, barley, Canada field peas, millet and corn. In the mash the meat scraps are fed four times a week. Oyster shell, grit and charcoal are always before the fowls. All mashes are seasoned with salt. Alfalfa is put in the mash every day. The above is the feeding for the commercial fowls. The breeders are fed differently during the hatching time and summer.

—Importance of Exercise.—

Exercise and good feeding are necessary to the well-being of the fowl. In the state of nature fowls have to forage for nearly all the food they get; and well fed hens as kept by those who feed for results the case is different and the heavy feeding tends to make the fowls heavy and sluggish. The proper digestion and assimilation of food depends on the proper amount of exercise. It must be compulsory and two of the three feeds that are fed in the day should be of grain and fed in a deep litter. Fowls can be compelled to take too much exercise; all that is necessary is enough to keep the digestion sound and the blood in good circulation, especially during the cold of the winter morning. This is one reason why we favor a grain food for the morning meal, although many use a mash in the morning with good results. If the mash is fed in the early morning, and the fowls should be fed as soon as it is light enough for them to see to eat, after eating the mash which is the bulk food of the day and the cheapest they will at once proceed to loaf the greater part of the morning. When the weather is very cold many after feeding mash scatter grain in the litter, but Biddy after her heavy meal does not feel like exerting herself in the effort to get the grain and this extra meal also calls for more work. We feed grain in the morning, which compels the fowls to work and thus they go until well towards noon singing and scratching and keeping warm and comfortable from the exercise. At noon we feed the warm mash and as the warmest part of the day follows the noon hour a little loafing will be less harmful than at any other time. At night corn is fed that they may go to roost feeling satisfied and warm. We feel sure that by using good judgment with the feed pail, by fighting lice, and with good houses and a good strain of birds that any earnest, hard-working person can make a success of the business by following the suggestions that we have made herewith.

GROWING WHITE WYANDOTTES ON FARM OF ARTHUR G. DUSTON, SO. FRAMINGHAM, MASS.

REGARDING WINTER EGG PRODUCTION

British Methods of Feeding, Breeding and Housing to Increase Winter Egg Production—American Breeds in England.

By H. DeCOURCY, Johnstown, County Kilkenny, Ireland.

THE egg is a natural spring and summer product and it is mere child's play to produce them in abundance during these balmy seasons, but winter egg production is another matter and calls for the exercise of the greatest intelligence, care and forethought that the egg farmer is capable of bringing to bear on his business. Winter egg raising is indeed a science in itself, and one that not one poultry keeper in every ten has succeeded in mastering. The reason of this is that there are so many little things which must be taken into consideration and which are apt to be overlooked by all but the keenest observers. In this article it is my purpose to put before readers of this book in the plainest and most unmistakable language a complete review of all the things which I have found by experience to be necessary for the profitable raising of winter eggs.

—Should We Raise Winter Eggs?—

This article is not intended for the eye of the fancier or the breeder of exhibition stock, since he has no business producing eggs in winter, because it pays him better to reserve the energies of his hens for spring laying, but it is intended to be a help to the "egg farmer" to the man (or woman) who makes his or her living by selling eggs for table use, or who aims at increasing his income by the pursuit of this industry. It is of the utmost importance to such person that they should understand how to raise a large proportion of eggs during the season, commencing November 1 and ending at the end of February, and the two most important reasons why this is so are these: (1) Because during the period which I have indicated eggs are scarce, dear and in very great demand, because the cream of the profit is to be made at this time, and whether the balance is to be on the profit or the loss side of the account at the end of the year depends very much on the number of eggs which the hens can be induced to lay from November to February inclusive. (2) Because every egg raiser supplies or ought to supply a certain number of private customers, such as families, hotels, restaurants, boarding schools, etc., etc., and the greater proportion of eggs he can sell to such customers the greater his profits are likely to be, but in order to be in a position to cater to this demand he must keep up the supply at all seasons—winter summer, spring and autumn alike. There may be other reasons why eggs should be produced in winter, but these are the two principal ones, and the practical poultry farmer ought not to lose sight of them. But now to come to something practical. Let us see how it can be done.

—How Shall We Produce Winter Eggs?—

The requirements for winter egg production may be briefly stated in the following words: (1) A good winter-laying breed must be kept; (2) the hens must be of a bred-to-lay strain; (3) the pullets must be hatched at the proper season—neither too early nor too late; (4) hens which are too old must not be kept in the flock; (5) all birds must be maintained in perfect health; (6) the fowls must be fed in such a manner and with such foods as are calculated to promote egg-production and (7) the houses, sheds and yards must be arranged so as to afford comfort and to insure perfect health. Having thus briefly stated the important things which the egg-farmer must bear in mind if he wishes an abundant supply of eggs in winter, my next business is to take up these items one by one as I have set them forth and deal with the most important points in each.

Breed.—It is quite impossible to single out any one breed and say: "This breed is a better winter layer than any other;" for the circumstances which suit one breed

Type of good winter layer.   Type of good winter layer.   Type of good winter layer.
Buff Orpington pullet.   Buff P. Rock hen.   Buff Orpington hen.

and bring it to the front as a good winter layer may be totally unsuitable for another. One man succeeds with one breed whilst a second "bets his last dollar" on another breed, and so it will be to the end of time. It is, however, possible to make a broader classification and when this is done the weight of evidence points to the conclusion that that class of fowls known as "general purpose breeds" is superior as a winter layer to that other broad class—the "table breeds" or even to the great laying and non-sitting class. The sixteen weeks' winter laying competitions which have been held in England within the past few years point to this conclusion and it is also my own experience as well as that of many other poultry keepers in these islands, who have kept records of the egg-production of various breeds of fowls, that the sitting breeds make the best winter layers.

When kept under favorable conditions any one of the following breeds can be depended upon to yield a good supply of eggs during the winter months: The Wyandotte (Golden, Silver, White, Buff, Partridge, etc.); Plymouth Rock (Barred, Buff or White); Orpington (Black, Buff or

White); Langshans (Black or White); and the Faverolles. I have never tested the laying powers of these in competition with such breeds as the Leghorn, Minorca and Hamburg, but others who have done so and who are to my knowledge careful experimentalists have assured me that when kept under similar conditions the Wyandottes, Plymouth Rocks and Orpingtons have proved far better winter layers than non-sitting breeds. This, however, is hardly a fair test, for it may be that the Leghorns or Minorcas could have done better under conditions more favorable to them. It is a matter which is very hard to decide and indeed the point can hardly be settled so long as different families of fowls require different treatment to bring into prominence their most useful characteristics.

For my own part I have always found the American breeds, or such of them as I have kept—the Barred, White and Buff Plymouth Rock and the White, Silver and Golden Wyandotte—first-class winter layers and more than capable of holding their own against the best of our English breeds. During three recent winters I have tested the laying powers of some of these breeds and I have kept careful records of their work, and the results are shown in the following tables of figures which I have prepared from my record books for the information of readers of the Inland Poultry Journal. I may say, however, that the American breeds used were not themselves imported birds, but were only the third generation from imported stock and were therefore fully acclimated.

Year—Month.	Av. No. eggs laid by Barred Rock hens	Av. No. eggs laid by White Rock hens	Av. No. eggs laid by W. Wyand'te hens	Av. No. eggs laid by B. Orp'ton hens	Av. No. eggs laid by Buff Orp'ton hens	Av. No. eggs laid by B. Langshan hens
1901—November	9	9	8	10	7	6
1901—December	12	14	15	12	11	9
1902—January	13	14	9	8	14	10
1902—February	15	18	12	14	15	16
1902—November	8	9	7	10	9	5
1902—December	13	11	14	11	11	9
1903—January	11	15	10	10	13	12
1903—February	16	17	14	12	13	17
1903—November	10	11	10	8	9	11
1903—December	11	11	12	13	13	8
1904—January	14	16	10	13	16	12
1904—February	16	19	16	11	12	13
Totals for sixteen months of three winters	148	164	137	132	143	128

The above are the averages for each individual hen and as there were thirty hens in each lot and the experiments extended over a period of three winters the results may be taken as a fair indication of the laying powers of each breed. I am aware, however, that the strain has much to say in the matter, but I intend to refer to the question of building up a good winter egg-producing strain later on. The age of all the hens was one and a half years at the commencement of the trials, and they were therefore just a good age for winter laying. During the same years I made experiments with six lots of pullets of the same breeds as those mentioned above, but I have not space to give the records here; nor is it necessary that I should do so, as the results are substantially the same as those recorded above. The White Plymouth Rocks led the way with the greatest number of eggs and were followed by others as shown in the table except that the White Wyandottes did a little better and made almost as good a record as the Black Orpingtons, which came third in the list.

Strain—I believe that the tendency or capacity to lay in winter depends more on the strain than on the breed, and that it is possible to build up a hereditary strain of winter layers, just as it is possible to breed a strain of deep-milking cows or of fast-trotting horses, or of sweet-singing canaries or of musical frogs. To accomplish the desired result the trap-nest must be brought into use especially during the winter months, for the hens which make the biggest records for the year round are frequently not the best winter layers. Those hens which lay most regularly and frequently during the three or four hard months beginning with November are worth their weight in gold or at any rate in silver, and they should be specially marked and reserved for the breeding pens of the following spring. These should also be mated with cockerels, whose dams and granddams are known to have been good winter layers, for the introduction of blood through the male side from a strain of inferior layers will spoil the effects of many years labor. On the other hand much can be done toward increasing the winter egg supply by persistently breeding from males and females of good winter laying strains, and also by the selection of the best eggs—those of good size and shape, strong shell and of the color which is most in demand, be it brown or white.

The true fancier, who breeds his birds for excellence of shape, plumage and other points which will win him honors in the shown room, pays very great attention to the question of strain, and will not introduce fresh blood amongst his flock without the deepest consideration and the most careful inquiry into the ancestry of the birds which he intends to introduce as breeders to improve the size or stamina of his flocks, but unfortunately the raiser of market poultry and eggs does not pay the same attention to the importance of building up a strain suitable for his purpose. Within the past year or two, however, I have observed that the raisers of farm poultry are beginning to mend their ways and to pay more attention to the strain of their fowls, and it is probable that they are led on to do this by the example of American breeders, whose work is chronicled in the various journals which circulate in the British Islands.

—Watching and Rearing the Pullets at the Proper Season.—

The egg-farmer can never make any decided progress in building up a strain of winter layers unless he breeds his own pullets to replace the old stock which he sells off every fall, for pullets which are picked up here and there can not be depended upon to lay in winter; but if he has his own breeding pens and facilities for hatching and rearing the required number of chicks he is master of the situation. In this connection I may say that very much depends upon the time at which the pullets are hatched and that there is such a thing as hatching too early as well as such a thing as hatching too late. Experience teaches us the proper time to hatch pullets of the particular breed which we keep in order that they may start laying neither too early nor yet too late, but to those who have not much experience in the matter the following advice and particulars should prove helpful. If we want our pullets to start laying at, say the beginning of November, and if we know the average ages at which pullets of different breeds usually commence to lay we can arrange to have them hatched at the proper time.

The following shows the average ages at which pullets of various breeds ordinarily commence to lay:

Class I—Hamburgs, Leghorns, Anconas, 4 1-2 to 5 1-2 months.

Class II—Minorcas, Redcaps, Houdans, Polish, Lakenvelders, 5 to 6 months.

Class III—Wyandottes, White Wonders, Plymouth Rocks, Rhode Island Reds, Orpingtons, Faverolles, Silver Gray Dorkings, 6 to 7 months.

Class IV—Langshans, Brahmas, Cochins, Dark Dorkings, 7 to 8 months.

Thus we see that if we want to have pullets of class I laying in November it is early enough to hatch them in late May or early June, and the danger is that if they are hatched in, say February or March, they will start laying in August and will be in full moult just when they are most wanted to lay the eggs that bring the creamy prices. On the other hand, breeds of Class III and IV must be hatched early, say in March, in order that they may be sufficiently mature to commence laying before severe winter weather sets in, for otherwise it is likely that they will postpone the job until the balmy days of spring, and no feeding or care can induce them to start earlier. Very much depends upon the care bestowed upon the birds. They must be bred from healthy, vigorous stock, must be well fed and kept growing by the use of wholesome foods, calculated to build up healthy, robust bodies, from the day they leave the shell until they have come to maturity.

HOW TO MAKE POULTRY PAY.

—The Age of Hens for Winter Laying.—

Every poultry keeper who keeps hens for the production of eggs for market aims at obtaining the greatest possible profit out of his flock by keeping the birds as long as they are profitable egg producers, and then disposing of them to the best possible advantage, but in these islands as well as in America practical poultry keepers hold widely different opinions as to the number of years hens should be kept, in order to make the greatest profit "in the long run."

By one set of poultrymen the system is advocated of hatching the pullets early, starting them to lay at the beginning of November, forcing them to lay right through the winter, spring and following summer by the use of highly concentrated egg-forcing foods, and disposing of them in the fall just before the moult sets in, when they are less than one and one-half years old. By this system it is confidently asserted that the cream of the profits is to be realized, but I believe that such a system, if persisted in must end in disaster. Here are the objections which I raise to it: (1) The labor is very great, and considerable accommodation is required to raise enough pullets every year to replace the entire stock of layers; (2) the continual hatching from the eggs of pullets which have not quite reached maturity must after some years result in the production of birds which are far from robust, although I agree if mated with two-year-old cocks, no evils may arise from hatching the eggs of pullets fo. one, two or three years. It is the long continuance of the practice of breeding from immature stock which must inevitably have an injurious effect; (3) if we sell off all our hens at one and the a half years old, it is impossible for us to utilize the trap nest in building up a strain of first-rate layers, since the utility of the trap nest in building up a strain of first-rate layers consists in ascertaining the first year the hens which are worthy of being retained the second and third years.

I believe that all practical poultrymen agree that is a bad paying business to retain hens for laying after they have attained the age of two and a half years, and it is indeed a fact that they are quite useless as winter egg producers at the age mentioned. Of course, if a hen has proved herself to be far above the average in her first and second years it is advisable to retain her in the breeding pen for three or even four years, so that she may become the mother of a large number of the young stock and may transmit her laying powers to her offspring, but speaking of the flock as a whole I am strongly inclined to the belief that the hen has done her most profitable work at the age of two and a half years and that is the time to get rid of her. When this is done, it is of course necessary to replace 50 per cent. of the flock every fall, and this can be done satisfactorily by raising the required number of pullets, since pullets which are brought up at random can not be depended upon as winter layers and frequently turn out most disappointing.

I have also referred to the labor and expense which are entailed by the raising of a large number of pullets every year, but this is an unavoidable expenditure and an eminent poultry expert, for whose writings I have a very high esteem, looks at the question through rose-colored spectacles and says: "If we raise fifty pullets we are forced to raise at the same time an equal or greater number of cockerels and we can sell the latter at a price that will pay for the raising of the entire lot of males and females. Then we shall have the price which we can obtain for the old hens as clear profit." Not a bad way, by any means, of looking at the question is that. I suppose all things have a bright as well as a dark side and everything depends upon the manner in which they are regarded. That which one man points out as a source of labor, trouble and expense is regarded by another as a source of gain.

1. Showing style of colony house converted into house and scratching shed for winter use.
2. Bird's-eye-view of the pens in which the experiments recorded by Mr. DeCourcy were made.
3. Photo: First Prize Breeding Pen of Buff Orpingtons, Cork International Exhibition 1908.

This subject brings me back again, incidentally, to the all-important question of "breed," which I dealt with and dismissed in a few paragraphs above, but must take up again for a few minutes' further discussion. The egg-farmer must depend not alone on the sales of his egg profit, but also on the sale of surplus cockerels and of old hens, and for this reason it is necessary that he should take into consideration the table qualities as well as the laying powers of his birds. I have already pointed out that the "general purpose" breeds are generally granted to be the best winter layers, but may add here that their fine table qualities must also be reckoned in their favor, since sooner or later practically every fowl raised on the farm finds its end on the table of the consumer.

Practical men will therefore do well to consider and study closely the demands of their markets for table poultry, and to keep a breed which comes near to fulfilling the requirements of the trade. I am well aware that in America the demand is for fowls possessing rich yellow legs and skin, but, as I believe most American poultrymen are aware from their intercourse with England, quite the opposite is required on this side of the Atlantic. Our epicures want and must get a fowl with white legs, white skin and white flesh, and will turn up their noses at anything dark or yellow. This accounts for the popularity of such breeds as the Buff Orpingtons, the Faverolles, the Houdan and the Dorking (especially the two first named), and were it not for the yellow legs and skin of the Rocks, Wyandottes and other American breeds I have no doubt that they would have long since become ten times more popular than they are in these islands. Their sterling qualities for general purposes are fully recognized, but there is a decided prejudice against them for the reasons I have mentioned, and since the introduction of the Buff and White Orpingtons several years ago the American breeds have steadily declined in popular favor. They could easily compete with and defeat the Dorkings, but it seems that they are not able to hold their own against the Buff Orpingtons. I believe that the American Rocks and Wyandottes are quite as good as if not better than the English Orpingtons in every respect, but in the scale with the Orpington we find the weight of prejudice and that pulls it down.

—The Importance of Health.—

This is a subject that, properly speaking, should have an article to itself, and it is not possible to deal with it adequately in the body of an article on another subject. Yet it can not be ignored, for this article would be very incomplete without a few paragraphs regarding a matter upon the whole fabric of profitable poultry culture depends.

The hen, in a state of nature, would lay and hatch only a few clutches of eggs in a year—perhaps only one or two—but the domestic hen of the present day is an artificial producer, built up by the efforts of countless generations of poultry keepers as a perfect egg-producing machine. We must, however, bear in mind that the hen is working at an unnaturally high tension all the time and that it takes a remarkably strong constitution to pull through all the work which she has to perform. To start with, therefore, we must be particularly careful in breeding from mature, robust and perfectly healthy stock, possessing no taint of hereditary disease. Our care in this direction must be followed by the utmost discretion in raising the chickens which are intended for next winter's layers, in healthy, open-air quarters, feeding them with the most suitable foods for promoting continuous growth of bone, muscle, flesh and feather, and protecting them against the danger of contracting any disease which may tend to enervate the system and render them unfit for the strain and exertion of the great work which they are expected to perform, when they come to an age to get into harness. Then, when the pullets have been moved into winter quarters and their period of usefulness has begun we must still be watchful and supply the foods which are required not only for the formations of eggs but also for maintaining the body in perfect health and replacing waste tissue.

In order that any animal may do its best work it is necessary that it should be fed with the food most suitable for generating the power to perform the allotted task. The horse, at the plough, in the hunting field or on the race course, will not perform his best work unless he has been previously fed upon the foods which give him the power and render him fit to do that work, and if his owner does not understand his business—does not know exactly what foods are required by the animal to perform that function—it is almost impossible that he can succeed in attaining his desired end. With other animals quite the same rule holds good. The cow will not yield a maximum flow of milk of the best quality unless she is fed with the foods containing the necessary elements for the production of rich milk and to any other animal, which one might mention, quite the same rule may be applied. The hen is no exception to this rule. If we wish her to produce an abundance of eggs it is obvious that we must supply her with the foods which we have found by experience to be the best for promoting the egg-yield.

Feeding for winter egg-production and to promote laying in summer are two widely different things, but just now it will be sufficient for us to consider the former since the subject is seasonable and the subject of summer egg-production can be dealt with another time. It is safe to say that a hundred different rations have been tried and found satisfactory by as many practical poultrymen, but the formula which I am now about to put before my readers is one which is largely used by poultry keepers in the British islands and which is also known to many in America, and I believe that it can hardly be improved upon although it may be varied to suit the individual taste or circumstances. Although it is now more largely used perhaps in Britain than in America, I believe that it had its origin in the latter country and its value has been strongly indorsed by Prof. Brigham, of the Rhode Island Agricultural College, who pronounced it: "A very well balanced ration, probably as near perfect as could be made with the material available on the average farm."

One strong point in favor of the ration is that it is almost entirely composed of foods which can be grown on the average farm, and since so many "general farmers" are now engaged in poultry culture, it is well that they should as far as possible utilize those foods which they grow on their own farms. I have not yet had a favorable opportunity of testing the efficacy of this combination of foods, but hope to do so the present winter. However, I have received very good accounts of it from neighboring egg-farmers and the following is the reply of one of them to my inquiries on the subject:

"Broadly speaking, the basis of our method of feeding is mash in the morning and whole corn at night, but as requested I have pleasure in giving you detailed particulars which you may include in your article:

Five mornings in every week we feed a mash, which we make in the following way: Fifty pounds of vegetables are cooked and mashed fine. This is placed in a large wooden vessel and mixed with one hundred pounds of mixed meal. Boiled milk or water is then added and the whole thing is stirred until it forms a stiff mash. As a condiment we add a teaspoonful of salt, another of ginger, a third of cayenne pepper, and a fourth of powdered charcoal to every bucketful of the mash and these are stirred in and mixed well. The mixed meals which are used consist of: Equal parts of middlings, bran, ground oats, barley meal, corn meal and beef scraps, measured in bulk by a scoop and thoroughly mixed before being added to the vegetables. We have generally a few pounds of table scraps every day and these are added to the mash just for the purpose of avoiding waste and because we can not put them to better use.

A very important constituent of the mash is the cooked vegetables which form its foundation. For this purpose we prefer clover to anything else, but we are not averse to using anything in the line of vegetables which may be grown on the farm and economically used as food for poultry. Such vegetables as potatoes, beetroot, cabbage, turnips, etc., are boiled before being mixed with the mash, and cut clover is treated in the same way. The mash which is made up in the manner described is fed for breakfast five days in the week and on Sundays and Wednesdays we feed whole meat for the morning meal. When the weather is fairly mild the hens have liberty to roam over a large field and they are fed only twice a day, the evening meal being of wheat, oats, barley and any other grain crop which may be available. We feed as great a variety of grain as possible. But when the weather is so severe that it is best to confine the hens to the scratching sheds, a light meal of wheat or oats is given them early in the forenoon and is buried in the litter so that they may be kept busy and not mope all day for want of employment. We have tried this method of feeding our laying hens during the past three or four winters and can strongly recommend it. Needless to say the birds must also be supplied with the usual accessories, such as fresh water, sharp grit, ground oyster shells, etc."

This is only one ration and there are dozens of others which may be recommended and are recommended in the American poultry press, but the main points to be remembered in preparing a winter egg-ration are that as great a variety as possible must be fed and that the foods must be largely of a nitrogenous character, consisting of milk, meat and vegetables, such as fowls feed largely on in spring and summer when they are allowed to run at large.

The American system of scratching sheds has within recent years been largely adopted by British poultry keepers, who are now fully alive to the importance of providing adequate day shelter as well as comfortable housing by night. Our winters are, however, comparatively mild and we have not to incur the same expense in erecting elaborate scratching sheds that American poultrymen must incur. A favorite plan is to convert the summer colony houses into shelter sheds by raising them two and a half feet off the ground and placing scratching material underneath. The space thus made is then sheltered by boarding on three sides and leaving one side open. This, however, is nothing more than a makeshift and I believe that ultimately the American system of scratching sheds will be adopted in its entirety.

PROFITABLE EGG FARMING *

One of the Safest and Surest Methods of Making Money with Poultry Is a Well-Conducted Egg Farm.

PRODUCING "strictly-fresh" eggs for market is recognized as the safest if not the most profitable branch of poultry farming. Certainly it is the easiest to master, and indeed the one line of poultry work into which one grows naturally. Of all branches of poultry keeping, egg farming carries with it the least risk from losses and the most certainty of immediate profit with quick returns. The successful egg farmer of to-day depends almost wholly on artificial incubation and brooding in obtaining his yearly supply of new stock birds. On many egg farms non-sitting varieties, chiefly the White Leghorns, are bred exclusively. On other farms the American varieties, particularly the Barred and White Plymouth Rocks, White Wyandottes and Rhode Island Reds are kept for egg production only, broody birds being broken up as soon as they exhibit a desire to sit.

During the past five years several of our State Experiment Stations have made careful investigation with regard to breeding for increased egg yield. These experiments have shown that as a rule pullets outlay year-old hens. Also that pullets which make a good egg record during their first year prove prolific layers during their second year and have the power to impart to their offspring the ability to produce eggs in large numbers. The best layers have invariably proved to be strong, healthy, vigorous birds. In experiments made by the Maine Experiment Station, covering a period of four years, in which breeding stock was selected by the use of trap nests, tests were given to over 1,000 hens, and among them thirty-five were found that yielded from 200 to 251 eggs each in a year. It was also found that a number of hens were poor layers, and some produced no eggs at all. The results of these experiments have been confirmed by many breeders who have adopted the pedigree method. In a case of breeders where careful selection has been the rule for a number of years, the progeny shows a large percentage of heavy layers. It will be at once apparent that it is necessary to adopt some method of getting rid of the drones, and securing the services of the best layers as breeders, for while it is true that some heavy layers produce pullets that are poor layers, they also produce a fair percentage of prolific offspring. On the other hand it has not yet been shown that a poor layer will yield good layers in the progeny. The selection of the male bird is an important factor in breeding for egg production. In every case he should be the son of a prolific egg producer, and good results will be more certain if his sire was bred from heavy laying stock.

Throughout the Eastern and New England States, notably in New York and New Jersey, are many farms devoted chiefly to egg production. On these farms the White Leghorn is the variety most popular, the two chief reasons for this being that they are considered to be the best layers, and because New York city and other prominent egg markets, outside of New England, prefer and pay a premium for a white-shelled egg. Boston, on the contrary, prefers a brown-shelled egg and pays a premium for it.

A noted egg farm is the White Leghorn Poultry Yards, Waterville, N. Y., which is planned to carry 5,000 head of laying stock and raises thousands of chicks annually. The favorite plan of house used here is 112 feet long, 16 feet wide, with a feed room at one end, and a walk (or alley) 3 feet wide in the rear of the pens. There are eight breeding pens to each house, 12x13 feet in size.

Another type of egg farm is one with semi-detached houses with fairly large yards (in which are set fruit trees), such as the one developed by Mr. C. H. Wyckoff at Groton, N. Y. On this farm the houses are 12x40 feet in size and divided equally by a partition through the middle. Flocks of fifty White Leghorn hens or pullets are kept in each pen and its corresponding yard or park. As the yards are 33x84 feet in size, the plan here adopted admits of housing and yarding 600 birds to the acre. When fowls are kept on this extensive plan it is essential that green food be regularly and abundantly supplied. This need has led to the preparation of alfalfa in several forms, offered for sale in 100-pound and 50-pound bags.

It is well known that there is a regular rise and fall in the prices of eggs, which are highest about Thanksgiving to Christmas, and lowest about May and June; the high prices of early winter being due to the scarcity of eggs at that time, the scarcity being caused by the old hens not having recovered from the effects of the molt and the bulk of the pullets not having begun to lay. It is plain that the greatest profit from eggs is made by producing them at the time of high prices, and this is the more obvious since the producing of eggs at that time hardly lessens the quantity produced at the time of lower prices; in other words, if the birds are got to laying in October they can easily be kept laying through the winter, and then will keep on laying of their own account; they lay in the spring time because "it is their nature to." March is the best month for hatching the American and April the best month for hatching the Mediterranean varieties when fall laying is desired, and with the use of the incubators we can time the hatches to the desired date. Six months of good, steady growth will bring Plymouth Rock and Wyandotte pullets to laying maturity, and pullets of these varieties hatched in March or April, if fed for growth, will begin laying in October and in November, December, etc., will be shelling out a full quota of eggs; similarly with Leghorns and Minorcas hatched in April and May, and if this has been done the profits of the high prices of Thanksgiving to Christmas, etc., will be the reward.

Pullets for Layers.

It is the pullets that do the late fall and early winter laying, and if we would have an abundant supply of eggs at the time of highest prices a goodly proportion of our birds must be early-hatched and well-grown pullets. An experiment illustrating the doubled profits to be secured from pullets as compared with year-old hens was carried on at the Utah Experiment Station, Logan, Utah, a few years ago, and the results were most instructive. Two pens of old hens averaged eighty-five eggs apiece, while two pens of pullets averaged 170 eggs apiece—exactly double the number. The doubled number of eggs laid by the pullets does not rightly present the ratio of profit, however, because from one-half to two-thirds of the increase comes at a time of decidedly higher prices, hence the proportion of profit is much greater. In this Utah experiment the average value of the eggs per hen was $0.78, while that of the eggs per pullet was $1.78. The food, cost, labor and interest on buildings, etc., is no greater for a pullet than it is for a hen, and it would seem to be the part of wisdom to have the bulk of our flocks early-hatched and well-grown pullets in order to secure our share of the profits from egg production. It is impossible to get a large number of early hatched pullets by the use of sitting hens. Hens sit when they please; incubators hatch when you please.

It ought to be noted that there is a good premium paid in the city markets for strictly-fresh-laid eggs, which decidedly increases the profits of the egg farmer, and this premium is paid for eggs that are marketed every day or two, or, at the very latest, twice a week; of these "strictly-fresh" eggs there is never an over-supply excepting in the spring of the year when everybody's hens are laying. We all want the goodly profits that come from successful egg-farming, and we can share in them if we will set our incubators so the chickens will be hatched in time to bring the pullets to full maturity by natural growth in October, and then it is an easy and decidedly pleasant task to keep them laying by good care and good food.

---

* Courtesy Cyphers' Incubator Co., Buffalo, N. Y.

HOW TO MAKE POULTRY PAY.

Fowls have always been considered a necessary adjunct to the farm. Years ago they did not look upon them with an idea of making money out of them; they were simply a necessary part of farm life, existing as much to eat up the refuse from the table and kitchen as for any other purpose. They laid eggs or not as it pleased them, and whether the production was more or less mattered little. No stock or crop on the farm received as little attention as did the fowls. They received little or no care, roosted in any place they found convenient, whether an old shed or in the trees, and lived chiefly on what they could pick up. In many cases they were lucky to be able to live at all, without exerting themselves to lay eggs. The question of age was never thought of; fowls might die of old age or disease and it was a matter of little or no consequence. Whether they molted late or early did not matter. The eggs were gathered at odd times and were often of questionable repute when obtained. Late in the spring sufficient eggs could usually be gathered to place under a broody hen, and the housewife would give a little time from her busy hours to growing a small flock of chicks. When the young male birds were large enough to kill late in the fall, they would be sent to market, rather to get rid of them than with an idea of making money. In spite of all this the farm poultry paid sufficiently well to keep down the store bill. This example of poultry on the farm, as it was in times past and is to-day on some farms, is not overdrawn. Such conditions might have existed indefinitely. The introduction of modern, up-to-date poultry papers, and modern methods in artificial hatching and brooding, has wrought a change and created a greater interest in the poultry business. The modern incubators and brooders have made it possible to hatch and raise chicks with the greatest ease and economy, so that the poultry industry is becoming one of the largest industries of our vast and growing country.

To be a successful breeder of poultry, whether for eggs, for market or for the show-room, one must be a successful feeder. Chicks die in the shell and out of the shell, and the uninformed person, earnest in his efforts to succeed, wonders why. It would be erroneous to say that unsuitable food or careless feeding alone is to blame for the trouble, but our experience tells us that taking the chick from the time it is hatched until it is three months old, half the deaths, roughly speaking, may be attributed to errors in feeding, the other half to the ravages of lice and the lack of suitable brooding accommodation.

If You Want Eggs, Feed for Eggs.

Stall feeding of cattle for market is entirely different from feeding for milk. The foods are dissimilar and the quantities differ. The object in one case is to secure meat and fat, in the other milk; and we all know that generally

View on the Light Brahma Farm of Frank P. Johnson, Sta. A, Indianapolis, Ind.

speaking the best milker appears to be little more than a bag of bones—her food is forming milk, not flesh.

Not only is it important to use foods suited to the purpose, but these foods should be the best obtainable. The dairyman that gets the best results from his cows feeds the best food. The trainer of race horses does not whimper over the cost of oats, but aims to get the last ounce out of his high-spirited runners or trotters when they enter the race track. The last ounce tells in eggs and fowls, and here too the quality of foods makes all the difference imaginable. It should be the object of every breeder, first to get the best foods, then he should lay himself out to get the most out of the foods by feeding them scientifically, either by mixing them himself in proper proportion or purchasing them ready mixed and reliable.

For egg production the object is generally to force the pullets to maturity, and after maturity until they are two-year-old hens, subsequent to which time they become less profitable. Laying foods pay for themselves many times over during the season of high prices for eggs, whether they are purchased separately and mixed by the poultryman, or bought in the prepared form. The proportion of animal food must be largely increased for this purpose, and a full supply of all foods should be fed. Do not overlook the bulky vegetable food—that is, the "green food," as it is generally called. It assists the fowl in digesting the additional food she is induced to eat when the poultryman is anxious to increase his egg supply.

Among the most useful books dealing with the subject of poultry culture in all its branches are Books 1, 2, 3, 4 and 5 of the Cyphers Series on Practical Poultry Keeping. Book No. 4 of this series is devoted exclusively to profitable egg farming. If laying stock are fed entirely on mash food or grain food in the open, it will produce an inactive fowl, which if it has a liberal supply of food will quickly become an over-fat one. A certain amount of fat is necessary to the good health of the fowl, but there is a wide difference between a bird that is in good condition and one that is over-fat. Exercise is of vital importance to insure good digestion and promote health. Breeding birds and layers cannot be kept sufficiently vigorous and strong by any method of feeding that will not provide exercise.

Briefly, the Cyphers method of feeding layers is as follows: Litter the scratching rooms or pens from eight to twelve inches deep with straw, hay or other litter material. In this, feed Cyphers Scratching Food twice a day, preferably as the first food in the morning and a little more at noon to keep the birds busy. This scratching food is a well-balanced ration that will aid the mash food in producing a healthy bird and a normal egg. It contains particles sufficiently small to promote exercise, and at the same time avoids waste of food, since the food contains no particle of grain that is too small to be found in the litter by the hen. With this food scattered in the litter morning and noon the birds are kept busy all day. The last thing in the afternoon just before roosting time they should have all they will eat up clean and quickly of a crumbly mash made from Cyphers Laying Food. This is a highly palatable ground meal mixture, mixed with the proper proportion of shredded alfalfa and pure meat food.

The quarters of the layers should be kept scrupulously clean, and the birds themselves free from lice. Keep grit and oyster shell before them in a self-feeding box at all times, and supply pure fresh water daily.

BUFF WYANDOTTES.

First cockerel and first hen at Chicago, also first cock at World's Fair. Bred and exhibited by Ackley & Page. Now owned by Simon Beuth, German Valley, Illinois.

BARRED ROCKS.

First cock and first hen at Chicago, December, 1904. Bred, owned and exhibited by by R. E. Hager & Co., Algonquin, Illinois.

HOW TO GROW CHICKS

The Real Facts Boiled Down—Experience of One Who Has
Made a Success—His Methods Described.

FOR several years the members of the Inland Poultry Journal Company have been making trips to an excellent fishing resort at Burlington, Wis., known as Brown's Lake. The nimrods in all sections of America are familiar with this resort, as it is one of the best bass lakes in this country.

There are a number of swell resorts on its border and owing to its location, less than one hundred miles from Chicago, there is a large attendance from May 25th, the opening of the bass season in Wisconsin, until late in September.

The most popular resort on Brown's Lake is owned and operated by Mr. C. W. Hockings, who was born and raised on the lake's border. His resort is beautifully situated back about one hundred yards from the water line with a natural park frontage down to the boat house.

For home comforts were everything to make an outing what it really should be, something to look back to with pleasure and to hope for its return, we have never found one that quite equals the Hockings resort.

Owing to the demand for fresh spring chickens as well as fish, Mr. Hockings found it a hard proposition to supply his table during the rush season, especially so when the farmers were depended upon to furnish them, as the late springs in that section of the State held back the average growth of broilers until late in July. The refrigerator, or cold storage chickens did not suit the Hockings family, and they are one of the few hotel families that believe their boarders should have as good as they desire themselves.

So a consultation was held as to how to supply the shortage, and they finally decided to raise the chickens themselves. Good grade stock was purchased to furnish the eggs, while incubators and brooders were purchased to rear the stock. Here is where Mr. Hockings went up against the hardest proposition he had yet tackled.

The incubator catalogues all told him how easy it was to raise chickens by the thousand, but when he began to put their advice into practice he discovered there was a screw loose somewhere. They either did not understand him or he didn't understand them. At any rate, the results were far from satisfactory and he decided to do a little work on his own hook and await results.

He made his brooders over, making them nearly three times as large as the one sent out by the manufacturers; he also made radical changes in the kind and system of feeding recommended by these companies. As soon as he began to do things his own way the chickens began to grow and the loss was brought down to a minimum—in fact, the loss was so small that others began to sit up and take notice.

We being directly interested in this line gave this part of his resort the most careful attention, and at no time or place have we ever found such healthy, thrifty chicks as we found here. He would tell us that we had to change our system of feeding if we ever hoped for success. When he showed us year after year flocks of young birds brought to broiler size with a loss not to exceed three per cent. we asked him to tell us in his own way just how he accomplished these results. After much persuasion and a long wait we finally got the information and we give it here in his own words.

Whether the same good results will follow in all cases and under all conditions we are not prepared to say, but we do know they have succeeded with Mr. Hockings:

First Meal

should be well baked white bread about a week old wet with sweet milk and squeezed out until it crumbles nicely. Feed only twice the first day, about 9:30 a. m. and 4:00 p. m., and keep fresh water before them all the time, from the first time you feed. The best way to feed the first two or three times is to spread a dark cloth over the floor in exercising room and put all the chicks on it, then scatter feed and they will at once begin to eat. Only leave them out ten minutes, then put them all back in nursery and shut them in and put drinking fountains in with them; take up cloth and shake it and it will be ready for next time. Don't ever leave feed before them; if you do you can expect trouble. At 4:00 p. m. feed again, same way, and fresh water every time with the chill off. It is a good practice to scald out drinking fountains at least every other day. Run hover temperature in cold weather at 90 to 95 degrees, and in warm weather 85 to 90 degrees, reducing to 80 as soon as chicks are two weeks old.

Second Day Feeding.

Feed three times—morning, noon and night—bread and milk the same as first day, putting them back in nursery each time and shutting them in; fresh water every time you feed. Most of them have learned to eat by this time. Don't leave any feed before them.

Third Day Feeding.

Feed four times—morning, 10:30 a. m, 2:30 p. m. and at night. Now they are getting lively and most of them will come into the exercising room themselves if you knock on the floor with your fingers. Eight minutes will be long enough to leave feed before them now, as they eat faster. You can leave the door to the exercising room open during middle of day now, shutting them in at night again. Of course, in real warm weather you don't have to be so particular in shutting them in the nursery. The main thing is not to chill them or get them too warm either. Now this is the day for their first grit. Scatter a little before them just before you feed at 10:30 so that each one can get only a few particles. Be careful, not too much grit, but any sharp, coarse sand will do. Feed nothing but good bread wet with milk (or water occasionally instead of milk is about as good), always squeezing out until it crumbles.

Fourth Day Feeding.

Now that they are crazy to eat be careful. Feed four times a day from now on; fill them full but don't leave any feed before them. Give them one feed of grit a day from now until they are two weeks, when you can leave grit before them all the time. Don't forget to scald out drinking fountains.

Fifth Day Feeding.

Feed bread and milk three times and for their night feed give them the meal mixture, wet with water and squeezed out the same as the bread so it will crumble nicely. Don't forget grit and fresh water, always cleaning up feed should you happen to throw down more than they will clean up in eight minutes.

Meal Mixture.

Fifty per cent. corn meal, 25 per cent. shorts, 25 per cent. bran and 10 per cent. beef scrap by weight.

(Use yellow corn ground fine and sweet bran and shorts fresh from mill.)

Sixth Day Feeding.

Feed bread in the morning, meal mixture at 10:30, bread at 2:30 and meal mixture at night. Now you can let them out of the brooder into the room during the middle of the day if it isn't too cold. Have the floor covered with litter; barn chaff is best. They know the difference now between dirt and feed. You can also scatter a little pinhead oat meal in the litter and teach them to scratch. Have litter about one inch deep and a very small handful of oat meal is plenty for the first few days.

Seventh Day Feeding.

Feed meal mixture in morning, bread and milk at 10:30, meal mixture at 2:30 and at night, not forgetting grit and fresh water.

Eighth Day Feeding.

Now they are what I call "on to their feed." Feed the meal mixture four times a day from now on, as regular as possible, and a feed of bread occasionally instead of meal mixture is good for them. Clean the brooder at least every other day after they are four days old; fill and trim brooder lamp every day, always being careful that it is put back right and not turned too high. Always call your chicks when you go to feed them, and always feed near brooder when outside, then if a storm comes up suddenly you can always call them in quickly. If at any time your chicks get dumpish and don't care to eat, just let them go without one feed; they will be ready to eat next time all right. Don't show your friends how nice they eat unless they are there at feeding time. Don't use any patent chick feed. Don't feed sour feed; a small handful of feed at a time. Feed only at feeding time, no matter how many times you go to look at them. They are crazy for four or five days; temptation's great. Danger lies in feeding too often or leaving feed before them.

Get your brooder ready two or three days before hatch is coming off. Bed brooder with clean straw cut four to six inches long; no chaff or grit in brooder for first week. Run hover temperature in cold weather 90 to 95 degrees; in warm weather from 85 to 90 degrees, reducing to 80 degrees as soon as chicks are two weeks old. Put chicks in brooder the next morning after hatch is over and don't feed until the oldest chicks are sixty hours old.

Do not keep your eggs in cold cellar. Keep them in ordinary living rom. Do not let your eggs chill at any time.

Brooder must be thoroughly warmed up before putting chicks in it, in early spring especially.

C. W. Hockings.

EGGS, BROILERS AND ROASTERS *

An Article in Which the Writer Has the Courage to Present Actual Figures of What Costs, Expenses and Profits Should Be on a Properly Conducted Poultry Farm.

By CHARLES A. CYPHERS, Buffalo, N. Y.

POULTRY AND EGGS are produced by those who do not aim for a large surplus above their family needs; by those who plan for the surplus to produce some little income; and by those who make a sole business of producing for the market. The principal surplus of the first class is eggs at a time when every one's hens are laying their best, when the markets are well supplied and the price is at the lowest. From the chickens raised each year a small surplus is sold, usually with little thought as to the grade of fowl or the time for disposal.

Of the second class, the production of a surplus of eggs during the high-price period of fall and early winter is the chief aim. The profits in the high-price period of fall and early winter is the chief aim. The profits in the proper marketing of the surplus cockerels is given little attention.

The third class is as yet very much in the minority, although its numbers have multiplied a hundred fold during the past ten years. This class runs to some special branch of the business. One works chiefly for market eggs; another for poultry meat, either broilers, roasters, or ducklings; and others for fancy or pure-bred stock and eggs for breeding purposes.

The great majority of the first class are farmers, small or large, and were it not for the intervening factor of the cold storage, the profits accruing to this class would not be worth mentioning. Poultry, like other animals, has a natural season for reproduction. To produce the egg, and from the egg the broiler or roaster, out of the natural breeding season requires especial care and feeding not found in the curriculum of the average farmer in caring for his flock of chickens. The consequence is that a great mass of eggs and poultry is produced during the natural breeding and growing season. Time was when eggs during these spring months dropped to five and six cents a dozen in the rural districts, and the average run of poultry thrown on the market during the late summer and fall brought correspondingly low prices. During the late winter and early spring fresh chickens brought premium prices.

The advent of cold storage made it possible to absorb the surplus during the time of plenty and preserve it, to be placed on the market during the time when the fresh article could not be so readily produced. To-day eggs command sixteen to eighteen cents a dozen even at the height of the laying season, owing to the sharp competition of the buyers for cold storage purposes. Prime roasters also bring premium prices for this purpose, while the miscellaneous cull stuff brings a fair price.

The rapid increase in the use of artificial methods has produced a large amount of fresh stock during the off season, so to speak, but the rapid increase in the population of the country, and especially the rapid growth of our large cities, has so increased the consumption and the demand that the increased production, together with the large tonnage put in the freezers every fall and marketed during the winter, has not decreased the price. The only effect has been to make the season of high prices somewhat shorter, but during the past five years there has been little or no change in the length of the season. The increased consumption has kept pace with the production, and with the settling up of the prairie lands and the cutting up of the large ranches, the possibility of low-priced beef and mutton has decreased, and there is little likelihood that we shall ever again see low-priced eggs and table poultry, or much change in the average run of prices for the different seasons of the year. We can, therefore, lay out our plans for poultry raising on the basis of current conditions, with a reasonable assurance that these conditions will be maintained for an almost indefinite period.

—Weight and Condition.—

To market a bird as a prime broiler or a prime roaster is to market it when in the pink of condition. This condition may be at two separate stages of its growth, or only one, according to the natural weight of the breed, and the sex. I will mention certain breeds as typifying given weights, and the peculiarities of those weights. I will use the Leghorn as typifying the light weight egg and broiler breeds; the Wyandotte the light weight and the Plymouth Rock the heavy weight general utility fowl; and the Light Brahma as the extra heavy roaster or capon type.

In the Leghorn class a pullet is in the pink of condition at one pound, and will remain in a plump condition up to one and one-half pounds. A cockerel is only in this condition at the latter weight. Beyond this weight they both go to frame. Birds of this class mature early, when they begin to lay or tread. The flesh then rapidly deteriorates in quality, becoming hard and stringy. Such a bird can no longer be marketed as a prime chicken, but it goes into the fowl class. It is, therefore, most profitable to market this class of birds as broilers. Some breeders, by selection, have materially increased the average weight of their birds, but a bird in the Leghorn class would hardly be selected for a roasting chicken.

In the Wyandotte class are the medium weight varieties, the cockerels of which are in the pink of condition at one and one-half to two pounds, and ripen at about four pounds, after which the flesh begins to harden up. The pullets of this class make as fine a light-weight broiler as the Leghorn and hold the plump condition up to one and one-half pounds, and are in good form at two pounds. They are usually marketed as broilers if not intended for breeders.

In the Plymouth Rock class the cockerels ripen at five to six pounds and the pullets at four. A Rock cockerel should not be marketed as a broiler, as at one to two pounds it would be all frame. When growing soft winter roasters, the pullets are often marketed at two pounds.

By caponizing the Rock cockerels may be grown to eight pounds. Above that the weight is taken on too slowly to make it profitable. As a bird approaches its maximum weight the more food is required to put on a pound of flesh. A caponized bird makes rapid growth within certain limits and will attain a large size without the flesh hardening. The larger the breed the more cheaply can a heavy weight be produced, and for this reason the Light Brahma is usually selected for growing heavy capons.

—The Basis of the Price Table.—

In making up the price table, I have used the wholesale prices paid by the commission men of New York City. I have chosen that city, as New York is practically the controlling market for the larger portion, based on population, of the United States. The Boston market prices sometimes rule a little higher for a short season; the Philadelphia, Baltimore and Washington prices a little lower, and those of Chicago and the central western cities lower still, owing to population and location in relation to the supply as furnished by the contiguous territory.

Aside from the general market of our large cities, with their congested population, special markets can be found the country over. Wherever people congregate at either summer or winter resorts will be found a demand for fresh poultry and eggs at premium prices. The poultryman must study the possibilities of his own location. While eggs and poultry can be profitably shipped a long distance by the producer, the large packers, with their numerous buying stations in the States farthest from the

* Reprinted by courtesy of Mr. Cyphers from his booklet: "Eggs, Broilers and Roasters."

HOW TO MAKE POULTRY PAY.

large markets, with their chains of refrigerator cars, bring the market to the farmer's door. If the farmer's eggs are fresh and the poultry meat of prime grade, with the same grade that he would furnish in a well-fattened hog or steer, he can usually command a good price, for, unlike the hog or steer, the poultry and eggs can be marketed direct by express at a good profit.

The tables are based on the average prices for the past three years, for strictly fresh eggs and prime poultry meat. Those who take no pains to market their eggs while fresh or who produce the eastern "cull" or the western "scalawag" cannot expect to secure these prices. Much of the poultry and eggs that come to our great markets seem to be graded only with a view to escaping the condemnation of the health commissioners. Many times the market is glutted with this inferior stuff that "just happened," while prime stock is searched for at premium prices. In fact, at almost no season is the market well supplied with the strictly fresh prime article.

EGG AND POULTRY PRICES.

—Averaged from January, 1903, to November, 1905.—

For convenience each month is divided into four even weeks or periods, and the four squares in each bracket, on the line of the month, shows the average price paid during the four consecutive weeks of that month.

stock is relied upon to supply the demand for roasting chickens, bringing twenty-one to twenty-two cents per pound wholesale. By the last of March the farm-raised stock is too old and hard, and when it comes on the market is sold as fowl at twelve to fourteen cents. The fresh chickens that find their way in during this period are produced by a few "soft roaster men" who begin to hatch in August and make a business of growing soft winter roasters. These men make large profits, but are as yet so few that their total output no more than supplies a special fancy trade, and these birds are not as yet found on the open market.

—Basis of Costs.—

I have figured the cost for eggs on the basis of two eggs for each bird raised to marketable age. The average cost to produce an egg is one cent. My estimate is based on the cost. If a breeder has a good egg market and wishes to know whether he could better afford to market the eggs or grow broilers or roasters, he has only to deduct two cents from the selling price of two eggs and deduct the balance from the profit as shown in the table for a chick hatched three weeks later. For instance, eggs are bringing two cents each the first week in September and the breeder is running a broiler and egg farm. If incubated, these eggs would hatch the last week in September, and the bird hatched would be marketed the third

Month	Hen Eggs				Duck Eggs				1-lb. Broilers				1½ to 2-lb. Broilers				4 to 6-lb. Roasters				8-lb. Capons			
	1st W'k	2d W'k	3d W'k	4th W'k	1st W'k	2d W'k	3d W'k	4th W'k	1st W'k	2d W'k	3d W'k	4th W'k	1st W'k	2d W'k	3d W'k	4th W'k	1st W'k	2d W'k	3d W'k	4th W'k	1st W'k	2d W'k	3d W'k	4th W'k
Jan	32	30	29	30				30	35	35	35	35	23	23	24	25	18	18	19	20	20	20	20	20
Feb	29	28	30	23					35	35	35	35	25	25	25	25	20	20	20	20	21	21	21	21
March	23	18	17	17	32	36	32	33	35	35	35	40	25	27	27	32	19	18	18	18	21	21	21	21
April	18	17	17	18	31	31	28	26	40	40	40	40	35	35	37	37	20	20	23	24	21	21	24	25
May	18	18	18	18	23	22	22	20	50	47	47		40	40	40	44	25	25	25	27	30	30	30	32
June	18	19	19	19									44	38	32	31	28	28	28	25	35	35	35	30
July	19	19	21	21									29	28	27	23	25	25	20	20	30	30	24	22
August	21	22	23	23									22	21	21	20	20	20	20	20	22	22	22	22
Sept	24	24	24	24									20	20	21	21	20	19	22	22	22	22	22	22
Oct	25	25	25	26									22	22	22	23	21	20	20	19	22	23	24	22
Nov	29	30	33	33									23	22	23	23	18	17	18	18	20	20	20	20
Dec	32	32	34	33									23	23	23	23	18	18	18	18	20	20	20	20

—Analysis of the Price Table.—

The first bracket in the price table shows the market price for fresh eggs. The premium price paid to regular shippers of choice graded eggs is from three to five cents per dozen more. The New York market pays the premium price for the "selected white," and the Boston market for the "selected brown" eggs. The second bracket shows when the premium price is paid for duck eggs.

Extra plump broilers and roasters will bring a premium of from one to three cents above the regular market quotations. From the fourth week in January to the third week in May, one-pound broilers pay a premium over heavier birds, as shown in the third bracket. During the balance of the year one-pound birds are in fair supply and bring no more per pound than one and one-half and two-pound birds which net the grower a greater profit. The price of the one and one-half and two pound broilers is shown in the fourth bracket.

The price on squab broilers weighing nine to twelve ounces is not shown. There is no constant demand for these, but the commission men notify their regular shippers when wanted. They are used for banquets, luncheons, etc., and bring from eighty cents to a dollar a pair.

The fifth bracket shows the prices paid for heavy roasting chickens. These must be soft meated—that is, young virgin cockerels, capons, or pullets before they begin to lay. The sixth bracket shows the average price paid for capons.

The average price paid for heavy soft roasters and capons from the last of March to the last of August is not so well fixed as during the balance of the year. In fact, fresh stock of this class during this period is so scarce that no regular market quotation is made. By the last of March the fresh stock is nearly all cleaned up and frozen

week in December at one and one-half pounds, at a profit of sixteen cents. If the two eggs had been sold they would have brought two cents above the cost, and the broiler therefore shows a greater profit by fourteen cents.

In beginning business with purchased eggs it will be necessary to pay a premium for desirable stock, and the difference between two cents and the price paid for two eggs should be added to the cost.

The cost of making a broiler or a given weight roaster does not vary much with the different breeds, within the possibilities governed by the mature weight of the bird. That is, one could not make a six-pound soft roaster of a Leghorn cockerel because the average mature Leghorn is altogether too light in weight. On the other hand, a Leghorn will make a very handsome broiler weighing one to one and one-half pounds, and at the same approximate cost as it can be done with the Wyandotte. What advantage there is, is in favor of the bird that will make a given weight in the shortest space of time.

An eight to ten week chick, weighing one pound, will cost approximately as follows:

```
Eggs, two per chick, at 1 cent......$0.02
Labor ..............................  .02
Feed ...............................  .05
Fuel ...............................  .01
                                     -----
Total ..............................$0.10
```

In the following gross costs I have included the cost of dressing but not the expense of icing, packing or expressage. Icing and packing would only cost a small fraction of a cent per pound, and the expressage varies so much it could hardly be averaged. The breeder can easily ascertain the express rate to his shipping point and de-

HOW TO MAKE POULTRY PAY.

duct it from the net profit. The express companies do not charge on the total weight of the iced package but on the weight of poultry contained therein. The commission man's charge of 5 per cent. is deducted in the final table, in even cents, the fraction ignored or added as it amounted to less or more than a half cent. The weights given are the usual profitable weights to market.

Gross Cost, Marketed at 1 Pound—
    Cost of 1-lb. chick..................$0.10
    Dressing .... .03
        Total $0.13

Gross Cost, Marketed at 1½ Pounds—
    Cost of 1-lb. chick..................$0.10
    Feed, ½-lb. at 5 cents............... .025
    Labor .... .005
    Dressing .... .03
        Total $0.16

Gross Cost, Marketed at 5 Pounds—
    Cost of 1-lb. chick..................$0.10
    Feed, 3 lbs. at 6 cents............... .18
    Feed, 1 lb. at 7 cents................ .07
    Labor .... .04
    Dressing .... .05
        Total $0.44

Gross Cost, Marketed at 6 Pounds—
    Cost of 1-lb. chick..................$0.10
    Feed, 3 lbs. at 6 cents............... .18
    Feed, 2 lbs. at 7 cents............... .14
    Labor .... .05
    Dressing .... .05
        Total $0.52

Gross Cost, Marketed at 8 Pounds—
    Cost of 1-lb. chick..................$0.10
    Feed, 3 lbs. at 6 cents............... .18
    Feed, 2 lbs. at 7 cents............... .14
    Feed 2 lbs. at 9 cents............... .18
    Labor .... .07

THE KIND OF POULTRY THAT BRINGS THE PREMIUM PRICES.
Specially Fatted, Correctly Dressed Poultry as Displayed for Sale in a Fashionable Butcher Shop.

Gross Cost, Marketed at 2 Pounds—
    Cost of 1-lb. chick..................$0.10
    Feed, 1 lb. at 5 cents................ .05
    Labor .... .01
    Dressing .... .03
        Total $0.19

Gross Cost, Marketed at 4 Pounds—
    Cost of 1-lb. chick..................$0.10
    Feed, 3 lbs. at 6 cents............... .18
    Labor .... .03
    Dressing .... .04
        Total $0.35

    Dressing .... .05
        Total $0.72

Gross Cost, Marketed at 9 Pounds—
    Cost of 1-lb. chick..................$0.10
    Feed, 3 lbs. at 6 cents............... .18
    Feed, 2 lbs. at 7 cents............... .14
    Feed, 3 lbs. at 9 cents............... .27
    Labor .... .09
    Dressing .... .05
        Total $0.83

The following tables showing the period necessary to produce a given weight are based on the assumption that the breeding stock is of good vitality, and up to standard weight. One cannot make roasters of canary birds, and the breed must have other of the qualifications than the mere name. A Rock weighing no more than a Wyandotte must be reckoned as in the Wyandotte class, or a small Wyandotte in the Leghorn class.

—Weights Required of the Breeders.—

Leghorn class—Pullet, 3 pounds; cockerel, 4 pounds; hen, 4 pounds; cock, 5 pounds.

Wyandotte class—pullet, 5½ pounds; cockerel, 7½ pounds; hen, 6½ pounds; cock, 8½ pounds.

Rock class—Pullet, 6½ pounds; cockerel, 8 pounds; hen, 7½ pounds; cock, 9½ pounds.

Brahma class—Pullet, 8 pounds; cockerel, 10 pounds; hen, 9½ pounds; cock, 12 pounds.

—Time Necessary to Produce a Given Weight, the Bird as a Roaster to Arrive at Market Size Within the Softmeated Period.—

In the Leghorn Class—1 pound, 10 weeks; 1½ pounds, 12 weeks.

In the Wyandotte Class—1 pound, 8 weeks; 1½ pounds, ten weeks; 2 pounds, 12 weeks; 4-pound cockerel, 21 weeks.

In the Rock class—2 pounds, 10 weeks; 4-lb. pullet, 21 weeks; 5-lb. cockerel, 21 weeks; 6-lb. cockerel, 25 weeks; 8-lb. capon, 30 weeks.

In the Brahma Class—5-lb. pullet, 21 weeks; 9-lb. capon, 30 weeks.

—Selection of Breeders.—

In the selection of breeders, for either eggs or market poultry, strong vitality is the keynote of success. A highly profitable strain of birds is only made by careful selection of the breeders, in choosing for this purpose only perfectly developed, bright, healthy hens, and vigorous, active males. Given a good inherited constitution and an abundance of fresh air and plenty of the right kind of food, we have the foundation for a good laying strain of birds. If, in addition to eggs, market poultry is to be a factor in the output the culling cannot be done too closely. It should begin while the birds are young and growing. At this time the difference in development is most noticeable. The culling should continue up to the time of going into winter quarters, picking out the birds that do not make a uniform quick growth. After the birds are in winter quarters, keep an eye out for drones. Cull out the birds that fill up with food and then mope around a good part of the day. These birds are unprofitable to keep for egg production, are usually plump in body, and should be marketed. A careful selection for a few years and the tone of the flock as a whole is brought up and a high point of productiveness attained.

In selecting the male bird, only make use of the strong, vigorous specimen—the bird with a clear eye and alert carriage. It is always well to mate cockerels with hens and cocks with pullets. With such matings better average results can be attained.

In mating, one male to eighteen or twenty females in the Leghorn class is sufficient. In the heavier breeds one to fifteen is better. Do not mate up the birds until about two weeks before the eggs are wanted for hatching and then break up the matings as soon as the eggs are no longer wanted for this purpose. By mating up only a short time before the eggs are wanted for hatching a stronger germ will be secured. When breeding from yearling hens stronger stock will also be assured if they are given a rest after the moult in the fall and are not fed strongly for egg production until January. This, of course, relates to the yearling hens from the eggs of which we wish to hatch the next season's layers—birds we desire to make superior in form and vitality to their progenitors. If the poultry breeding has been begun with pullets or purchased eggs, and only pullets are available, select for the breeders only those that are well developed, and birds of uniform shape and size. By doing so a more uniform lot of youngsters can be secured, and progress will be made from the start.

In proportioning the number of breeders needed from which to secure the eggs to hatch a given number of layers for the following season, plan upon securing about forty eggs per hen during the natural breeding season. From these forty eggs at least twenty-five youngsters should be hatched and grown. Half of these will be cockerels, so that each breeder will not produce more than ten to twelve selected pullets for layers. This estimate is based on using all the eggs for twelve to fourteen weeks.

In selecting the breed of fowls one must be governed by whether the aim is to produce poultry meat or eggs as the main output. In selecting layers for market eggs, much depends upon the size and strain. That is, the small breeds are somewhat the most prolific, and, size and other conditions being equal, birds of a strain that has been built up by careful selection for vigor will be found the most profitable egg producers. The Leghorns are generally conceded to produce the largest number of eggs, but one can hardly say that any breed is pre-eminently the egg bird, as there are strains of Wyandottes, Rocks, Reds, Minorcas, and other breeds that are very prolific producers. The Light Brahma is seldom kept for market eggs, but almost exclusively for growing heavy roasters. The soft-roaster men will pay sixty cents a dozen for these eggs from August 1st until late in January.

In catering to the New York market, a large white egg, such as laid by the White Leghorn or Minorca, will command the premium for table purposes, although the hotels pay a premium for freshness rather than color. The Boston market prefers the rich brown egg, and throughout New England may be found "brown egg" strains of Wyandottes, Reds, Rocks, and other breeds. There is no difference in flavor between a brown and a white egg, but the latter can be more readily candled for blood spots, which will occasionally appear in the egg.

Birds of the Wyandotte class are as fine meated as the Leghorns, make plumper broilers, and can be grown to roaster size. Birds in the Rock class make the popular well-plumped soft roasters, weighing five to six pounds. The White Rocks are a little the heaviest of the breed, and on account of the yellow skin and freedom from dark pin feathers, with clean yellow shanks, makes a very popular roasting bird. The Orpington makes a large, fine-meated bird, and the quality of the flesh is fast overcoming the prejudice against the light skin and willow shanks.

—Which to Start With, Eggs or Hens?—

How the start shall be made will depend upon circumstances and the season of the year. If the beginning is made in the late summer or fall, and the laying houses can be made available, I would recommend purchasing breeders. If no laying houses are available, and the start is made during the winter, there is little choice but to purchase eggs. If the start is made in the spring, I would purchase eggs from which to hatch the next year's layers and breeders, and use colony brooders and small colony houses to rear the youngsters, and do no building until the following fall. There is little difference in the expense if good stock is secured, but what advantage there is in favor of purchasing the breeders. In either case it will pay to get good representative stock of your chosen breed. If you already have some stock that is not quite up to grade, the purchase of a good male bird will be found an economical way to quickly improve it. To start with the best one can afford and then by selection and an occasional purchase, bring the flock up to an even grade of excellence, is a good rule to follow.

—Housing the Breeders.—

When breeding for special markings it is generally best to mate up in single pens with one male. A house eight feet wide and ten feet deep is amply large for one pen of single mating. Such a house will comfortably care for eighteen or twenty birds of the Leghorn class, or fourteen or sixteen birds of the Rock class.

In breeding for market poultry, when only general type and vigor is important, large numbers may be most profitably kept in the same enclosure. On general principles the larger the flocks that can be kept together, the more profitably can the poultry business be carried on. A hen, as a unit, is a small item. It is giving consideration to each fowl that makes for a petty business. Make the unit a flock of fifty, or a hundred, or two hundred, and the work begins to take on the dignity of a business. It is therefore a good plan to construct the house so that birds in considerable numbers may be run together.

In moderate climates, where the birds can get out on the ground much of the time during the winter months, a house ten feet deep is ample. Where the winters are severe and there is much snow, it should be made at least sixteen feet deep. Even when using a continuous house it is a good plan to divide the rear half up into sections of eight, ten, or twelve feet. This breaks the draughts and makes the fowls more comfortable for the night. I prefer the eight or ten-foot division, as there is less disturbance at roosting time.

It is advantageous to use a platform beneath the perches to catch the droppings, as the fowls not only have the additional exercising room beneath during the day, but they more readily find a roosting place at night. An opening should be made in the partition at the end of the platform so that the fowl can go from one division to another until it finds a vacant space. The platform should be three feet above the floor and the perches twelve inches above the platform. Space the rear perches eleven inches from the wall and the others eighteen inches apart. The platform should extend eight inches beyond the front perch.

A good arrangement for the nests is to place them along the partition, with the approach on a level with the platform and the outer end closed with the exception of a half inch crack for light. The front may have a hinged door for gathering the eggs. In blind nests like these, above the floor the hens seldom acquire the egg-eating habit. If desired, the nests can be made double to accommodate a larger number of hens.

A division eight inches by sixteen inches with four perches will care for forty fowls of the Rock class or fifty of the Leghorn class. To mate these for breeders would require three males. To confine three males in this space would not work well. By placing the partition in only the rear half of the house, and leaving the front half open and continuous for several pens, the males can get away from one another and there is less disturbance. How many divisions may best be left continuous is still an open question. During the coming winter of 1905-6 I shall use forty-foot pens, with five divisions, holding 200 birds each.

Under the old methods of mash feeding large numbers were not conducive to the best results. Hopper-feeding methods have overcome any disadvantage of large numbers running together. It takes practically the same amount of time to give the necessary care to one enclosure, whether it contains fifteen birds or a hundred. Time is money to the man who makes use of it. It should not be forgotten that it is the numbers that count in the poultry business, and the facility with which these large units can be cared for.

The front of the house should be arranged so as to admit both light and air. Sunlight is a germ destroyer and fresh air and health go together. Some poultrymen prefer a closed house artificially warmed for winter work. Just enough heat is given to keep the house dry and free from frost. This requires careful manipulation to keep the fowls free from disease, and is wholly unnecessary. I prefer the partially open-front house. In such a house the hens are more hardy and there is less sickness. The type of house I employ, and which I here illustrate, I have used with the best success. In Buffalo, where we have much zero weather, snow, and wind, I closed the lower shutters only three nights last winter, and then it was probably unnecessary. In this house a pen of Barred Rocks gave me an average egg yield of fifty per cent. all winter. I not only left the shutters open but did not use the drop curtain at night. For Leghorns I should probably have closed the shutters, and perhaps used the curtain to protect their combs. At Lakewood with the mild Jersey winters, the curtain-front house is used altogether, but in northern New York State I prefer the glass in the upper half, with solid shutters below, hinged to swing out, to protect the house from storm.

Many of my customers write me regarding the advisability of using the colony house plan. I do not advise it, for the reason that one cannot care for as many fowls as with the continuous type of house. With the work concentrated one man can easily care for three thousand breeders, or five thousand layers for market eggs. If he has not this number that is no excuse for wasting the time running over a large acreage. There is no advantage in the colony plan, and much more work is entailed.

—Profitable Egg Farming.—

A study of the egg market and the natural conditions of egg production is a necessary preliminary to profitable egg farming. The table gives the average prices paid by the commission men of New York City for the past three years. These prices should not be taken as representing the prices paid by the hotels, fancy grocers, and butcher trade for strictly fresh graded stock. This trade habitually pays from five to twenty cents per dozen above the market price as made by the commission men, based on miscellaneous receipts—the greater the scarcity of strictly fresh stock the greater the premium. Moreover, all commission houses are willing to quote and pay a fixed premium of from three to five cents per dozen to steady shippers for eggs that they know to be strictly fresh and which they can so guarantee.

In order to secure the private trade or any special price from the dealer, one must have a sufficient supply to assure regular shipments, even if the amount is but a few cases a week. The common mistake of beginners is to look for the premium trade before they have the goods with which to supply it. The stewards of the large hotels will not contract for a great quantity with any one breeder. Individual shipments are too uncertain to depend upon. Some of these hotels use several thousand eggs each day, and it is no easy task for them to arrange to always have a supply of strictly fresh eggs sufficient to supply their needs. The better class of grocery and meat stores have a large trade that demands strictly fresh eggs; and to tell a customer at any time that these could not be supplied would mean the loss of that customer's trade. This class of trade does not ask the price but it does not listen to excuses. Even those in moderate circumstances who live in cities, where produce at best comes to them second hand, will not eat the stale article if a little more money will secure the fresh. Thus the commission house secures a premium for strictly fresh eggs and is willing to pay a premium.

From January 1st to July 1st, 1905, there were 2,225,709 cases of eggs marketed to New York City alone. This is 66,771,270 dozen, or 801,255,240 eggs. Still the commission houses paid the prices given in the table for miscellaneous lots of fresh stock. To secure a premium it is necessary to keep a large enough flock to insure regular shipments, that the receiver may come to rely upon the breeder making his shipments at regular periods. With such a shipper almost any large dealer will contract to pay a stated premium above the regular market quotations. Supply the fresh eggs in quantity and the premium price will be sure to follow.

In analyzing the egg table we see in the prices a true reflection of the natural productive and non-productive periods of the hen. The breeder who manipulates his flocks so as to secure its productiveness during the time of high prices is in the minority, and even should this individual multiply a thousandfold during the next ten years, this increase would not materially affect the market price. Too large a portion of the general supply comes from the miscellaneous small flocks of the general farmer, which are left to produce or not as they are influenced by the surrounding conditions. We therefore see the prices at the lowest during the early spring months when nature utilizes the storage of fat and energy of the unproductive period and forces the generative organs to reproduction.

As the spring advances many of the hens become broody and the egg production gradually falls off, so that by May we see a slight hardening of the market. During June and July there is a slight tendency to rising prices, so that by the end of August, when the hens are beginning to moult, the price is well on its way up. By the end of October the yearlings are about through laying and the pullets have hardly commenced, and during November, December and January we see the maximum prices prevail. The first touch of spring in February starts the prices on their downward course.

A hen will only lay about so many eggs in twelve months, and if the poultryman can influence this egg yield by careful management, so as to secure a substantial number during the time the usual farmer's flock is unproductive, it will bring the average price for the year to a highly profitable figure.

A well-selected and well-fed flock of pullets will average 150 eggs each in twelve months. The average farmer's flock will average about half this number. A hen that is laying heavily will consume more food than will a moderate layer. Near the large Eastern cities where grain is high the cost per annum to feed a fowl is from $1.20 to $1.30 per head. The average hen is ill fed, and if full fed it is usually with food that cannot be turned into egg material.

A bushel of corn will keep a hen a year, but on this ration she will not produce eggs in quantity, and the few that are produced are laid during the spring months when they are marketed at the lowest price.

A person who wishes to maintain a highly profitable flock of hens will therefore strive to secure layers that will produce when the average flock is idle, and in such numbers that a good profit on the investment will be made even at the lowest prices. Pullets of the Leghorn class begin to lay freely at five to six months, and those of the Rock class in from six to seven months. I have seen a few eggs dropped by Leghorns early in the fourth month and by light-weight Rocks early in the fifth, but profitable egg yields begin usually two months later. The early hatched pullets can be got laying in August, and if well fed the late hatched will commence to lay before cold weather sets in, when they will lay right through the winter. The birds that are laying strong by November first are the most profitable. Those that do not commence to lay until January only have two months of good prices and are doing their best during the height of the natural laying season—the season of low prices. They partially compensate for this by laying later in the fall, and should not be marketed before the houses are required for the next lot of pullets. If properly fed they will lay 30 per cent. during the moult, and usually will stop directly afterwards.

In housing layers for market eggs the same general principles apply as in housing breeders, with the exception that layers can be handled in larger flocks. With hopper-feeding methods pullets in flocks of five hundred can be made to produce as good an egg yield as in flocks of twenty, and the larger flock can be cared for with but little more effort than the smaller.

For market eggs no cockerels should be kept with the layers. It is the fertile egg that stales quickly, especially during the warm weather. The male is of no advantage in the flock of market egg layers, as they will lay quite as well or be.ter when no male is around. Keep the males by themselves and only mate the flocks up when the eggs are wanted for incubating.

Feeding layers for eggs should begin with the feeding of the chicks. In other words, to make the foundation for a profitable layer the bird must be well nourished from the time it is hatched until laying maturity. One of the most common complaints I receive in the fall is that the pullets do not lay. The reason usually is that they are not ready to lay because they are not fully developed. They have not been fully nourished and therefore run a month or two beyond the usual period of laying maturity. Hopper-feeding methods obviate this complaint. Where the pullet as a chick has been hopper fed with mixed grains and beef scrap, and has had all it wanted to eat, and when it wanted it, the chances are that it will begin to lay early and keep it up throughout the winter. When feeding at intervals it is difficult to feed all that is needed, and the chick does not make the best growth.

To make good winter layers, do not yard the pullets too early, but leave them in the colony houses on free range until late in October. They will then settle down to steady egg-laying shortly after being cooped up, and if full fed will lay well right through the cold weather. While in the colony houses do not shut them up closely at night, but rather let them get accustomed to the cold nights. This will harden them and help to fit them for the winter's work. It will be a kindness to the pullet and a money-maker beside. Always remember that fresh air is necessary to the best health and that, when the fowl is accustomed to it, cold does not mean discomfort.

In marketing eggs it will be profitable to assort them as to size and color. A uniform lot of eggs is always more salable than a mixed lot. If your fowls lay white eggs it will then only be necessary to cull out the over-large or small and deformed eggs. If of a colored variety, put the dark eggs together and the light ones in a separate compartment. If you mean to secure private trade, get a distinctive package holding a dozen eggs and bearing your name or the firm name, and seal it. See that these eggs are always fresh and your trade will increase and multiply and the premium asked can be gradually increased to a very profitable point. A customer lost on account of high prices is seldom lost for good if your eggs have been uniformly fresh, as you have the stale egg in the usual run of the market article to ever remind the rebellious customer that eggs under your brand are always fresh.

For select premium trade it is necessary to candle the eggs and take out any containing blood spots, which will occasionally appear in eggs from the best cared for flocks.

BLACK ORPINGTONS.

A drawing from life of the celebrated Black Orpington cock, "Duke of Kent," and three of his great sons. The "Duke" was bred by William Cook & Sons, of England, and has never been beaten in the show room in England or America. Mr. Cook and others pronounce him one of the most typical Orpingtons ever bred. He was bought at a fancy price in 1903 by Col. D. N. Foster, of Ft. Wayne, Ind., who still owns him and used him to found his "Duke of Kent" strain of all-purpose Orpingtons.

WATERFOWL CULTURE.

The Different Varieties and Their Relative Merits.

By R. J. HILL.

IN the East, especially around New York, or rather within easy shipping distance from it, are many large duck ranches that have passed years ago the experimental stage and are today paying propositions managed by business men on as sound a business basis as any other commercial enterprise. They are catering to the fast increasing demand for young or so-called "green duck" and at a price, all sides of the question being considered, that leaves a far better margin than broiler raising. In Epicurean knowledge the East is far ahead of the West. Not that they have better cooks or better things to eat, but simply because they pay more attention to the delights of good living. Green duck forms a part of every well appointed banquet during the season which extends over a greater part of the year. The time is not far distant when mammoth duck plants will spring up around all large Western cities that will possibly rival those of Long Island, where one ranch marketed 75,000 ducklings last year. There are some large plants in Pennsylvania built during the last few years catering to Philadelphia and I understand they are doing well, having a demand for their entire product.

There are several advantages in water-fowl culture over chicken raising. The housing proposition is very simple and inexpensive. The feed, while bulky, may be of a very cheap nature, especially for the breeders that are held over. One man can take care of a good many more ducks than he can of chickens. They are practically free from disease and last, but not least, the financial returns are quicker. There are openings today adjacent to Chicago, Cleveland, Indianapolis or any of the large Western cities for duck ranches, assuring the projectors good interest on the money invested.

Any city having a large foreign population is also a good market for geese, young or old, as long as they are fat. Any one wishing to take up waterfowl culture as a fancier will find an open field with very few competitors in this country. In England a fine Rouen commands as much attention as a good Barred Rock here and commands the "long green." It is true that waterfowl are not treated quite square by many show managements, they being placed in the dark part of the hall alongside of the pet coon, and the judge comes along after dark and gives first to the largest duck regardless of color or shape.

Varieties may be divided into the utility and ornamental classes. The first would contain Pekins, Aylesbury, Rouen, Blue Swedish, Cayuga and Indian Runner. The Pekin is conceded by all operators of large plants and by 90 per cent. of breeders at large as the high card in market ducks. They are the most desirable size, they dress nice, mature very early, lay a good number of eggs, as many as 150 a season, and are very hardy.

The Aylesbury is a pound heavier all around and matures slower than the Pekin. It is one of the prime market ducks of Europe.

The Rouen is also a popular European market duck and while of the same size as the Aylesbury, matures a little earlier, but the breed has no advantage to offset its color.

The Cayuga is one of the best of varieties, but its color is against it as a market breed and it will never gain a place as such. As a fancier's fowl it is a beauty and it takes patience and work to breed a flock without any white on breast or under bill. Great claims have been made for the Blue Swedish as a market breed, but they have failed to materialize. In fact, I do not know any one who is breeding them in quantities. As a fancy fowl they have not made many advances. The writer had the pleasure of placing the awards on largest display of this variety ever shown the year they were admitted to the Standard at Cleveland in 1902.

The writer also judged a small class the past winter in the East and must say they came nearer standard requirements, especially in wing and breast marking four years ago. Indian Runners are in a class by themselves and have considerable merit. They differ from other varieties in their decided upright carriage. They are small but splendid layers and are often called the Leghorns of the duck family. In the ornamental class I have placed the Muscovey. Some may object to this. While it is true some of our Southern brethren prefer this variety for the table the fact remains that their flesh has a flavor peculiar to this breed, and that many do not take kindly to it. Some claim that they are not a duck and should have a class to themselves, basing their claim on the well defined peculiarities of the breed. They have no quack, they fly as easily as a wild bird, they roost high and scratch worse than a cat. The period of incubation is thirty-five days. The "freakiness" of the bird is sufficient to make it a drawing card on the park pond. Either the colored or white variety is very pretty and I believe the White Muscovy is the most superlatively white thing in nature. The calls are the most popular of the domestic ornamental water fowl. The white is a cute and pert little thing, and the gray is very beautiful in coloring. The blending of the different shades of brown on the female is the most perfect harmonizing of colors to be found anywhere. The Black East Indian is another of the bantam ducks and it is a deplorable fact that we seldom see a good specimen in this country. Many of the birds showed as Indians are undersized Cayugas. I heard an exhibitor recently at one of the state fairs remark while East Indians were being passed upon that the principal difference between Cayugas and the modern East Indian is about two ounces. This is a breed for some true fancier to take up and push it where it belongs—in the front rank of exhibition waterfowl.

Breeding and Raising Market Ducks.

Every suburban resident is in position to raise at least a few ducks for the table. An accessible tub of water will take the place of the pond that was once considered an absolute necessity, and a small plot of grass for pasturage little later the number of females may still be increased per drake. Keep the flock confined until after eight o'clock and you will be able to find all the eggs without hunting. They drop them any old place. The eggs may be dirty; simply wash the dirt off, not the greasy coating. A good ration (morning) for breeders: Steamed cut clover 25 parts; corn, oat or barley chop, 25 parts; beans, 25 parts; middlings, 15 parts; gluten, 5 parts; tankage, 10 parts.

This is fed in deep trough, mixed rather soft. A hundred pounds of this mixture will require about 115 pounds of water. Night feed, steamed cut clover and bran, equal portions, and some whole corn, or cut vegetables (mangels, turnips, etc.), and whole corn. The amount of green food to be varied according to the available pasturage. If you use a machine for hatching use a good one—one that can be depended upon—then operate it carefully according to directions furnished by the manufacturers, air the eggs freely and the result will be good, strong ducklings if the parent stock is all right. If you use hens for hatching, select good, steady and proved mothers that may be depended upon to keep at her job four weeks, and above all have her free from lice. Ducks in their natural state do not have live and do not take kindly to them when they are forced upon them.

BREEDING DUCKS ON THE DUCK FARM OF JAMES RANKIN, SOUTH EASTON, MASS.

completes the natural surroundings required. The housing need not be expensive—a plain shed, storm proof and dry, with no internal fixtures, fills the bill as completely as the more elaborate house. The floor should be kept well bedded for sanitary reasons. A shed 8x16 feet gives ample area for twenty-five ducks and may be as low as the stature of the caretaker will permit. The yard should be as large as possible, though two hundred breeders may be kept on an acre of pasturage.

The larger the pasture the less feed will be required. Growing grass will furnish the greater part of a breeding duck's living and while they pick the growth close they do not destroy the roots as do chickens. Now in regard to the water question. As mentioned above, a pond or stream is not necessary, but either is a benefit to the breeding stock. Personally would prefer a swamp where the water is about six inches deep so they can work and feed at the bottom of the bunch grass and other semi-aquatic vegetation. It adds to the fertility of the eggs. The main thing to remember where the water is supplied is this: Have the utensils whatever it may be, of sufficient depth that the fowls can bathe their entire heads. Neglect of this precaution will result in sore eyes. Mate for early spring one drake to four hens, after April 1st one drake to five hens, and a

Do not feed the ducklings for from twenty-four to thirty-six hours. If you brood in a machine select a model that provides well for ventilation and carry the heat at ninety degrees, or a little above, the first three days and then gradually reduce. The first feed is an important one: Oat meal, ten parts; sifted bran, fifteen parts; white middlings, five parts; clover meal, ten parts; fine grit, two parts. Mix with hot water and allow to cool. Feed four times a day—better five times—on shallow tin plates. Wash the plates after each feeding. Use above ration for first week. For second week replace middlings with corn meal and add one pound of beef meal or scraps. Allow heat to drop to not less than eighty degrees. At end of second week and until the sixth feed framing ration: Corn meal, ten parts; oat chop, ten parts; bran, ten parts; clover meal, twenty parts; beef meal or tankage, five parts; grit, four parts. At six weeks of age the birds should have built a good frame. If for market confine in pens and feed corn meal bran and clover meal, equal parts. If to be kept for breeders allow free range, decrease the corn meal and feed more green food. The most profitable time to market ducks is just as soon as they are full fledged, threfore figure to have them in good flesh at that time. They dress easier and look nicer than at any other time. The duck is a very healthy bird and the saying is: "A sick duck is a dead duck." Generally speaking, they are subject to only two ailments —rheumatism and apoplexy. The first is caused by dirty or

THOROUGHBRED PEKINS.

wet quarters, and the latter (confined almost exclusively to brooder stock) is caused by absence of shade. Either of these can be guarded against.

Mating to Produce Exhibition Stock.

In Pekins select first birds with clean bills, then from these select birds with longest body and back. Do not go wild after big birds, but consider well the shape demanded by the Standard and remember that a bird of Standard weight is more apt to be typical than one two or three pounds over weight. In Aylesbury the principal points to be considered aside from color of bill is the horizontal carriage demanded in opposition to the partially erect one of the Pekins. In Rouen avoid breeders with white in wings. This will throw disqualified females every time. The same applies also to Gray Calls. In colored Muscovies, remember the blacker the better. In Cayuga and Black East Indians look for white feathers under bill of drake. In Indian Runners select all breeders from color. You will get enough gray specimens anyhow. While the Standard for the Blue Swedish calls for two white flights in wings, would prefer wings as dark as possible in the breeding yard.

At the beginning of this article I spoke of cheap food, and in enlarging on this phase of the subject will say that any vegetable, if cooked or cut fine, can be used, turnips, small potatoes, cow beets or even pumpkins. For animal food takage can be used instead of beef scraps. Ducks can not be fed off their feet like chickens can.

The market possibilities of geese are less than of ducks, but nevertheless the demand exceeds the supply. A limited amount of land will not, however, fill the requirements for goose culture. They need and must have, to produce profitably, plenty of pasture. It is said that ten geese will consume as much grass as a cow. The Standard recognizes seven varieties of diversified types. Toulouse, Embden, Brown and White China, African, Canadian and Egyptian. Of these the latter named two are strictly ornamental and can not be considered of any value as a market bird. The Toulouse and Embden are the largest, while the Chinas produce the largest number of eggs.

It is claimed that the African has the finest grained flesh, but personally I can not see any difference. All things considered, the Embden would make the finest market fowl. It dresses nicely and makes a good appearance. To be sure of fertile eggs breeders should be at least two years old and their usefulness lasts until they are ten years old or more. Sometimes a gander will only mate with one goose, sometimes with as many as five, but I believe one gander to two geese is the best mating. Confine the selected mating alone for a week or so and the probabilities are that they will stay mated for life.

Feed the breeders cut clover, vegetables and bran. Tankage can be added to force egg production. Best results will come by incubating under large hens rather than under geese and the period varies from thirty to thirty-three days. Generally speaking, a hatched goose lives as they are subject to hardly any disease. They must have shade and plenty of water. Twenty-four hours after hatching feed a mash of corn, oat and barley chop, five parts; bran, ten parts; tankage, one part. After the peepers are a week old increase the tankage. After three weeks of age they can pick their living from the pasture with only one feed of mash a day—at night. If incubator hatched do not commit the error of brooding in large flocks. Thirty in a bunch at the most. Twenty would be better. At four months of age they should be fit to kill and they do make good eating.

I do not believe there is a demand or a need for any new water fowl varieties. In shapes, sizes and colors we

A FLOCK OF INDIAN RUNNERS.

have sufficient to select from to enable the most fastidious to satisfy his desire whether for utility or fancy. There is room for improvement in the different varieties that will keep the best breedrs thinking to accomplish.

TURKEY REARING IN THE BRITISH ISLES

A Very Complete Article Telling of Methods Used on the Other Side of the Atlantic.—Turkey Hens Make the Best Hatchers and Mothers.

By M. DeCOURCY, Johnstown, County Kilkenny, Ireland.

THE busy season for the turkey breeder begins about the 1st of March, when the pens have been made up and the stock birds mated and when the first eggs of the season may be expected. The more advanced hens start laying at the time mentioned, and provided that the birds are in good condition for breeding the whole flock should be in full lay before the end of March. Laying will then continue throughout the months of April and May if the hens are not allowed to sit. Although the turkey hen unquestionably makes a better mother than either the ordinary fowl or the brooder, and although the best turkeys are those which are hatched early, it pays best to keep the turkeys from brooding until about the end of May, because turkeys eggs are exceedingly valuable and can be sold readily at from three to five dollars per dozen.

The Demand for Turkey Eggs

in all parts of the British Islands is almost always greater than the supply and a good healthy American Bronze or other purebred turkey will lay from three to four dozen eggs in a season, and will thus earn a gross profit of from ten to twenty dollars. I know of no kind of stock that can be kept on a farm that will yield as good a return

WELL BRED BRONZE TURKEYS.

as this; and as the cost of keeping turkeys should not exceed two dollars per annum, it is clear that cattle, sheep, pigs or any other farm stock cannot earn the same net profits. Many breeders will not sell eggs for hatching even at the high prices mentioned, as they consider it more profitable to hatch all the eggs laid, but in either case it does not pay to allow the hens to hatch too early in the season, and though they may go broody several times, it is easy to break them of broodiness by shutting them up in a barred coop and placing them where they can see the other turkeys on the farm.

The Laying Season

ought to be brought to an end about the end of May, as turkey eggs are of little value after that time, and it would injure the birds for next year's breeding to keep them laying too long. It is well that they should get a complete rest by setting them to hatch at this time, but as it is then too late for hatching young turkeys either hens' or ducks' eggs may be given to them. It will be found that they make excellent sitters and careful mothers, and indeed turkeys are frequently kept for the sole purpose of hatching eggs of various kinds.

In some parts of Ireland turkeys are very largely raised for the London and other English markets, and the industry is so profitable that many a farmer's wife depends almost entirely on her revenue from the turkeys to provide clothing for her family, and to furnish provisions for the household, when times are bad and the other takings of the farm are barely sufficient to pay the rent with rates, taxes and labor added. In the chief

Turkey Raising Centers of Ireland

the industry has been developed in a most businesslike way, and I believe that I can best convey a clear conception of the manner in which things are done by minutely describing the seasonable work on a successful farm. Suitable stock birds having been bred or otherwise secured, the breeding pens are made up early in the season and the nesting places are arranged in the following manner: A large barrel is taken and turned on its side, supported on trestles which raise it slightly off the ground and one end is knocked out. Several hens will make use of one nest for laying, but when the time for hatching comes each hen must have a barrel to herself. When the bird is to be set the barrel is thoroughly cleansed, all the old bedding is removed, the inside of it is limewashed afresh and a good nest of short dry straw is made. This is well dusted with insect powder and the turkey herself is treated in the same way. She is then put to sit on half a dozen "dummy" eggs, and the front of the barrel is closed by nailing over it a sheet of thin canvas to keep all other turkeys out. By this means a commodious, well ventilated nest is provided. In such a place the turkey can sit in comfort and seclusion, and there is not the same danger that she will break any of the eggs which there would be if she were confined in

A Badly Made, Ill Ventilated Nest.

After one or two days she will have taken to the dummy eggs and settled down to her work, and she may then be supplied with a full sitting of turkey, hen or duck eggs. A large turkey will cover seventeen or eighteen turkey eggs, but the usual brood is fifteen.

When all this has been done there is nothing requiring further attention except to dust the nest and the hatcher with insect powder once a week, and to feed the latter regularly once a day. A turkey hen is preferable to any other mother for young turkeys, because she cares for them better than an ordinary hen, and seems to know by instinct what is best for their health and welfare. But for the reasons already given turkey hens are not always used, and it is customary to employ large farmyard hens to assist them. A large Cochin, Brahma or Plymouth Rock hen can

Hatch Nine Turkey Eggs

and if a turkey lays thirty or forty eggs before becoming broody it is customary to get two or three ordinary hens to help her hatch the eggs. When the young pullets are leaving the shells the farmyard hen is also found most useful, for some turkeys are nervous and awkward, and if left to their own devices they crush the eggs or the young. When "the hatching out day" comes a good plan is to examine the eggs and place any which are chipped under the ordinary hen, and give those which are not chipped to the turkey. Then when the poults have come forth, and have become dry and hardy, they can be returned to the turkey, whilst the hen is taking care of another hatch of chipped eggs. Of course, this kind of work can also be done by an incubator if there is one on the farm, and indeed,

Incubators are Now Extensively Used

for hatching turkey eggs, the young birds being given to a turkey to be reared. During the time of hatching, if the weather is wet it is not usual to moisten the eggs, but if the weather is dry they are sprayed daily during the last week. Warm water is used and it is sprayed lightly over the eggs at feeding time, immediately before the hen is

replaced in the nest. The eggs are tested on the sixth or seventh day and any which are found to be infertile are removed. It is, however, quite unusual to find an infertile turkey egg, and provided the stock birds are healthy and properly mated, the eggs are always fertile. Should any infertile eggs be discovered in a hatch it is probable that all the eggs of the turkey which laid these few infertile ones will prove barren. In such a case there is something radically wrong with the turkey. As to

The Best Time to Hatch Turkeys

there is some difference of opinion, and whilst it is well known that the earliest birds thrive best, it must also be admitted that because of the uncertain spring weather which we have experienced within recent years, it may be expedient to defer hatching to a little later than the customary time. It is certain that the earliest hatches give a considerable amount of trouble in showery April weather, and taking all things into consideration I would advise turkey raisers to have patience and not to set the eggs too early. The last week in April is a good time for the poults to make their appearance, but if they are hatched out any time in May they are quite early enough and have plenty of time to grow and to attain the desired weight before Thanksgiving or Christmas. When the turkey hen is employed to

Hatch and Brood the Young

she must not be given full liberty to take them where she will. This would certainly be fatal for her instinct does not teach her that turkeys in a state of domestication are less hardy than wild turkeys. There is a fair share of the nature of her wild progenitors in the domestic turkey, and probably this is why she will start off at daybreak, when the grass is literally soaking with dew, and take her poults of two or three days old for a ramble. Young poults cannot stand a severe wetting nor can they stand the amount of exercise which the mother will give them if she is left to herself. They want hovering as much as anything during the first few days, but the mother seems determined to walk them off their legs.

A Useful Brood Coop.

In the British Islands generally, but more particularly in Ireland, the climate is very changeable, and in May we have more or less rain every day. For this reason it is necessary that young poults should be cooped, and a special turkey coop, which is largely used in Ireland is worthy of detailed description. The accompanying illustration serves to show the general outline of this coop, and the dimensions are: Eight feet long, four feet wide, four feet high to the apex, and three feet to the eaves. It is divided into two compartments by means of a wooden partition, made partly of laths, which are so placed that the young birds can pass from one division to the other. The sleeping compartment measures three feet by four, and here the mother is confined. The size of the other division is four feet by five and it may properly be called the run. The sleeping compartment is made of boards nailed closely together, but there is

A Free Passage of Air

through the partition, and this is quite ample since the sides of the run are made of wire netting, and do not interfere with the current. If there is one thing that will kill young turkeys faster than any other it is close cooping, and therefore the coop to which I refer is especially designed with a view to the admission of an ample quantity of fresh air. The whole of the combined coop and run is covered by a waterproof span roof, made of three-quarter-inch boards, and the boards which compose the sides and ends are of the same thickness. The bars between the two divisions of the coop are set four inches apart; the netting forming the sides is of one-inch mesh and the appliance is bottomless. The cost of all the material required to make it is

Not Above Two Dollars

and a carpenter can construct it in from one to two days.

It frequently happens that the duties of hatching and rearing young turkeys are entrusted to farmyard fowls of the larger kinds, and it is sometimes necessary to do this, although we know that the poults seldom grow so well or so large when mothered by a hen as they do when under the care of a turkey. The objection to the hen as a mother is that she will not take her brood far enough afield, but is inclined to keep them about the farmyard. She also makes a practice of hovering her chicks in the same place night after night until it becomes filthy, and this is a thing that the turkey will not do, for she seems to know instinctively that her brood wants a frequent change of ground and that it is expedient to change the roosting place often. Accordingly it is necessary that when

A Hen is Given the Care of Poults

she should be kept in a portable coop and should be moved to fresh ground daily. When this precaution is taken the ordinary hen raises poults very well; but perhaps the greatest drawback is that she forsakes them too soon—in fact, before they are half grown and at a time, too, when they need special care. On the whole, it is, therefore, ad-

AN AMERICAN BRONZE GOBBLER.

visable if possible to have a turkey hatching at the same time as two or three ordinary hens, and they can be allowed to assist her in raising the combined broods until the latter are several weeks old, and afterward the turkey alone will take charge of them. When newly hatched turkey poults are exceedingly tender, and need to be carefully treated. The methods of rearing them are indeed many and varied, and it goes without saying that there are more than one or two plans which may give

Equally Good Results.

Much depends upon the individual operator, I am inclined to think, and that system which suits one rearer may be a failure in the hands of another. But, however this may be, I shall describe the methods of rearing turkeys, which are followed in one of the most successful turkeys farms in Ireland.

The nests used are roomy and safe and when the shells have been taken away the poults are left unmolested to their mother's care until they are about twenty-four hours old. They are then taken from the nest and fed for the first time. The favorite plan, provided the weather is favorable, is to move the poults, with their mother, from the nest to a coop, which is placed in a short grass plot. The coop has a run attached in the manner already described, and in this the poults are fed.

The Food Consists of Custard

made from eggs beaten up with milk and mixed with finely chopped green stuff, such as leeks, onion tops or lettuce. It is also thickened by the addition of fine oatmeal or prepared chick meal. This food is crumbly and rather dry and is placed in small heaps on a piece of planed board, two feet long by four inches wide; and with much care and more patience the tender poults are given their first lesson. When the meal is over the board is scraped clean into a vessel, and the food left, if it has not been soiled, can be mixed with fresh food and used again. The poults eat very little at their first meal, but after an hour they are

Ready for Another Feed,

and want very little teaching this time. Now is the time to supply them with some fine, sharp chick grit, which, just before the second meal, is placed on the feeding board referred to. The poults do not eat it freely for the first time, but they soon learn to relish it, and henceforth a supply is put upon the feeding board, and the poults partake of it before every meal until they become quite accustomed to eating grit. It is afterward placed in a special grit box, and always remains in the run within their reach. The meals are given five or six times a day for a week, commencing with daybreak and ending with sunset, so that the interval between meals is only about three hours. From the commencement

Milk is Given the Poults to Drink

and it ought to be largely used for this purpose on all farms where it is plentiful. At this stage of the young turkey's life the most important matters which one must remember are: That cleanliness is absolutely essential to health; that the foods fed must be sound and wholesome; that they must be supplied regularly—a little at a time—and that plenty of green food, grit and milk is conducive to health. A small quantity of animal food is also fed in the form of fine beef scraps—only about five per cent. in the mash. The mother must not be forgotten but must be liberally fed on oats, corn, etc. For a week the poults are fed without change in the manner described and during this time they are confined entirely to the roomy coop and run. The coop is constantly moved about from place to place, so that every day the poults have fresh ground; and should the weather prove unfavorable they are confined for a second week or longer.

After the First Week

the weather being fine, the young turkeys are allowed to run with the mother for a few hours each day in a paddock which has short herbage and which is only a few acres in size, so that they cannot ramble too far afield. There is but little alteration in the foods or manner of feeding, but eggs are dropped from the bill of fare, and the mash now consists of a large proportion of equal parts ground oats and cornmeal and a small share of milk-curd, together with a large quantity of green food. Animal food is also discontinued, as the turkeys can find sufficient in the run in the form of insects of many kinds. Turkeys, young or old, seldom eat earth worms but they thrive on the slugs, beetles, grasshoppers and other insects, which are to be found in gardens, fields and orchards. Whilst it is necessary that

Young Turkeys Should be Protected

as far as possible against severe wettings, it is equally imperative that the birds should be allowed as much liberty on range as possible from the time they are six to eight weeks old, so that they may become robust and may have full opportunity of building up a gigantic frame, capable of holding the flesh which is later to be piled on in preparation for the great festivals of Thanksgiving and Christmas. During the growing period they thrive well on pasture, meadow or orchard, and when the corn and other grain has been removed off the stubbles the turkeys may with advantage be turned on to the grain fields to glean the ears and scattered grains that have escaped the hands of the harvesters.

LIGHT BRAHMAS.

Bred and owned by Frank P. Johnson, Station A, Indianapolis, Ind. The birds in the group won first prize as pen at the Great Indianapolis Show, February, 1906. The hen in the foreground has won six first prizes for Mr. Johnson. This hen and the one with the well spread tail were in first pen at the St. Louis World's Fair.

POULTRY HOUSES

How to Build—Location Has Much to Do with the Kind of Building—The Slanting Front House Preferable.

By CHAS. E. CRAM, Carey, Ohio.

A GREAT many things are to be taken into consideration in the construction of a poultry house, chief among which are those things looking to the welfare of the birds themselves that they may be healthy and productive.

Some things must be considered in some localities that are not thought of in others. A case in point came under our observation a couple of years ago. A poultryman who lives near a creek that occasionally becomes a raging torrent in the springtime, was showing us his poultry house, pointing out the manner in which he had constructed nests, roosts, dropping boards, etc., and then, pointing to a small attic in the gable of the house which was partially floored, he stated that in case of a sudden rise of the creek and he didn't have time to get his birds out they could fly up in the attic till the flood was over. Few of us have to contend with such difficulties, yet no doubt, the spring floods in many localities are a cause of great loss to poultrymen. A few years ago we received a letter countermanding an order, the writer stating that his poultry buildings had floated down the Miami and that he was completely discouraged. We may not all have to contend with floods, yet no doubt, many contend with some particular thing peculiar to their locality or their premises in the construction of their poultry houses, so that no one particular design of poultry house could be recommended as the one par excellence for all people and for all localities. Some general principles may be stated, however.

Poultry houses must be so constructed that the interior will be dry; they must have plenty of floor space, sufficient windows to admit an abundance of sunlight, and must be free from drafts. These things are absolutely necessary for the welfare of the fowls and the success of the poultryman. To be dry it must have a tight roof. Shingles, roofing paper or felt are to be preferred, as metal is excessively hot in summer where the sun shines upon it. The first cost of paper is less and it does very well if painted, but shingles in some localities at least, will be as cheap in the long run. The wall should preferably be of matched siding, but may be of common boards lined with paper on the outside or both. If lined on the inside it keeps out the drafts, but hardly keeps the walls dry. If lined on the outside the walls are kept dry. We think it is economy, however, to use matched siding. Then, if additional warmth is desired, it may be lined on the inside and the paper remain on the walls for a number of years. It would be still better lathed and plastered, but this is quite expensive, still many contend that it pays, especially in cold climates or with the small breeds. We have heard it said that plastered poultry houses were damp, but we can see no reason why it should be so, and it has not been so with the one house of that kind with which we are familiar. Dampness may occur in some plastered houses as it does in others, but it could probably be traced to a leaky roof or a damp floor. If there is any one thing more important than another about a poultry house it is that the floor must be dry. It may be made of dirt, cement or boards. The fowls do not care which, just so it is dry and clean. Our poultry houses are filled with dirt about 6 inches higher than the surrounding ground and they are dry all winter. There should be no trouble from damp floors in a poultry house so filled if in an ordinary well drained location. Some contend that tight board floors are better, but we have had no experience with them. We would not want a board floor with open cracks in it, but if it is tight we can see no reason why it should not be at least as well for the fowls and it may be easier to clean, and if we had to buy dirt with which to fill we believe we would buy flooring instead. Cement is more expensive and can hardly possess any advantage over either of the others except as a protection against rats, with which we are not troubled.

Drafts from cracks in the walls or under the roof, or from ventilators, especially if combined with dampness, have sent many a chicken to its eternal rest. Small cracks are as bad as large openings, especially if located near the roosts. Drafts and dampness in a moderately warm coop are worse than dry excessive cold or even roosting on the limb of a tree in the open air. Chickens, like persons, catch cold from sudden changes of temperature or from uneven temperature of the body as is the case when either are exposed to drafts. Catching cold isn't part of the process of freezing to death. As drafts then lead to colds and colds predispose to roup, the importance of preventing them is apparent. Ventilators are not an unusual source of drafts. In ninety-nine poultry houses out of every hundred no ventilators are needed. After we have made the house as tight as we can make it there are always enough openings around windows and doors to admit of sufficient ventilation in winter, and in summer we can leave the doors and windows open. If the poultry house is kept clean, one will not need to worry about ventilation.

Our own particular poultry houses were constructed with these ideas in mind and these objects in view. Two of them are of the slanting front design. They are 14x28 feet, divided into three pens each, 5 feet high on the lowest side and about 7½ feet at the highest point. In constructing them, posts are set in the ground at the corners and at two or three places along the sides, extending about 8 inches above the ground. To these were spiked a 2x8 plank standing on edge, which served as a platform. Matched siding was used all around. This was put on running up and down. In siding up the front, at intervals where windows were desired, we planed the tongue off of a board and in its place planed a groove with a plane made for the purpose, so that we had a groove on each side of the board. Then these were placed just far enough apart so that the glass would slide up and down the groove. This groove was unfortunately made a little large, so that the glass slide passed each other. To avoid this we nailed small braids along the glass to hold it in place. In summer we remove a few of the upper glass or slide them past each other to give a better circulation of air. This makes a cheap window frame and does quite well. The glass do not overlap, but come just edge to edge, and the leakage is very slight. This large area of glass and the slanting front makes a very warm house, especially when the sun is shining, and usually the first thing a stranger says when he enters the coop in winter is, "How warm it is in here." Have doors at both ends of the coop and roosts at the back. The buildings cost about $25 each, exclusive of the work. Have never had a frozen comb or wattle or a case of roup or other contagious disease in these houses. They were built two years ago. At that time we had seen one that was giving satisfaction, and heard that Chas. McClave was building slanting fronts to the exclusion of all others.

The advantages of a house so built is that a greater area of floor space is obtained at but little more expense than a smaller one with upright front, more warmth is obtained, as the sun's rays fall more directly into the coops, and it makes a drier floor for the same reason. Its principal disadvantage is that the front will leak unless it is well painted, for it partakes partially of the nature of a roof and must shed water. The prime coat of paint should be put on immediately after the boards are on or a rain and the hot sun will warp the boards and leakage result. This is where we made a mistake. Our coops were built in the fall. Painting was delayed until one rain after another prevented the work being done until spring. Another mistake we made was in not sufficiently supporting the front. Sometimes in winter great banks of snow are piled up on the front, and unless sufficiently supported to sustain this weight they settle, as mine have done, to a slight extent.

We have heard persons speak discouragingly of slanting fronts, but we believe that faulty construction is largely the cause of dissatisfaction. All poultry houses should preferably be built in the spring or summer, and if dirt floor is used filling should be done in September or before. If done after the fall rains begin, the floor is liable to be damp all winter.

COLONY HOUSES AND BROOD COOPS

W. A. Doolittle, Partridge Wyandotte Specialist, Sabetha, Kan., Gives Descriptions and Dimensions of Styles in Use on His Poultry Farm.

As illustrated herewith they are made 6x5 feet 4 inches, 4 feet high at back, 5 feet 4 inch in front with shed roof, as shown in picture, and are made as follows: For sills or runners I used 2x6 cypress, as cypress lasts very much longer when placed on the ground than either yellow or soft pine. The sills are cut six feet long with sled runners at each end. Then stand edgewise and tie across with three pieces of 2x4's cut 5 feet 4 inches laid flat and morticed into top of sills or runners, one at each end and one in the middle. The floor is made of 8-inch ship lap, cut 6 feet and nailed lengthwise of the frame on the 2x4's. The sides and ends are made of 8-inch ship lap running up and down, cut 4 feet for back and 5 feet 4 inches for front, and run down over frame two inches and is nailed. The top is nailed to a square frame 6x5 feet 4 inches, made from 2x3 sawed out of 2x6s. This completes the frame. The roof is made from 8-inch tongued and grooved soft pine, cut long enough to extend about six inches over at back and front.

The door which is simply 3 of the 8-inch ship lap, is at the front righthand corner of the end and is left down about two inches from top so that it can be swung open without interfering with roof. It is hinged to open out and back. Running up and down at each side of the door is a 2x3, to which a screen door is hinged so as to open in and to the front, made out of 1x6 fencing. The cross piece at the bottom is 1x6; top and side pieces are 1x3, ripped out of 1x6s. The screen doors and windows are covered from the inside with 1-inch diamond mesh poultry netting made from No. 18 wire, galvanized after it is woven. This is much stronger and better than the ordinary and makes the house rat and skunk proof. The screen doors are for the purpose of additional ventilation during the hot summer days and nights and may be fastened with a hinged hasp and locked with a small padlock.

TYPE OF COLONY HOUSE USED BY MR. DOOLITTLE.

lock. I use what is called the R. F. D. mailbox padlock. Each has an individual key that will not unlock the other, but there is furnished with them a major key which unlocks all.

In front is one 6-light 8x10 window hinged at the top so as to swing out and act as a storm protector when open; at the front righthand corner is a small trap door, hinged at the top with a spring screen door hinge which holds it open when raised and holds it shut when lowered. This trap door can be used for birds to pass through and main door or screen kept locked except when passing through to clean the house.

The object of these colony houses is to place twenty-five or thirty chicks in as soon as large enough to leave the brooder house and as they are on runners can be drawn with a horse to any part of the farm and colonized so that each little flock has its own range and if placed at

BROOD COOPS IN ORCHARD.

the edge of farm or pasture where insects are plentiful the results cannot help being most satisfactory, as no method of feeding can equal free range and plenty of insect life.

Coops for Hen and Chicks.

The illustration gives but a poor idea of the coops as they are placed in the orchard under the trees which furnish shade and protection from storms. However, they are most satisfactory and are made as follows:

Two pieces of 2x4s 3 feet long are placed edgewise on the ground. Use shiplap for flooring cut 4 feet long and 2 feet 6 inches wide. Nail this to the sills which raise the floor 4 inches from the ground and keep it dry in heavy rains. The sills project about two inches in front and four inches in the rear for coop to rest on, as it is made so as to fit over the outside of floor, and they carry the rain off and not to the floor as it would if made to rest on the floor.

Make a square frame for top and bottom out of 1x3s cut out of 1x6 fencing the proper size so that outside of frame is the same size or a trifle larger than the floor. To this frame nail the back and ends, running up and down, back 23 inches high and front 27 inches. This gives a shed roof and as stated allows the side and end boards to extend down over the floor and rest on the sills.

The roof is made of half-inch lumber put on crosswise of the coop and cut 4 feet 6 inches long so as to project out over the front of coop eighteen inches and about six inches at rear. This front projection protects front of coop from storms that would otherwise beat into the coop. The roof should be covered with some good roofing paper; I used 2-ply rubberoid. The whole front of coop excepting eight inches at righthand corner is covered with heavy galvanized screening, one-fourth to one-half inch mesh. The righthand corner above spoken of is used

HOW TO MAKE POULTRY PAY.

Breeding Pens and Runs.

Colony Houses Arranged in Orchard.

VIEWS ON MR. DOOLITTLE'S PARTRIDGE WYANDOTTE FARM.

for a doorway. A solid door is used for closing the coop at night and renders it rat and vermin proof. A slat door is made to take the place of the solid door when it is wished to retain the hen and allow the chicks to run out during the day.

The screen usually comes thirty inches in width and as there would be a 6-inch strip come off of it as the coop is not high enough I used this strip in the back by running the back boards up within about five inches of the roof and nailing the 6-inch strip above mentioned over it. This gives ventilation through the coops, but as it is at the top and over the chicks they are not harmed by the draft.

You will see I have not fastened the coop to the floor, but placed it over the floor, resting it on the sills which, as stated, project beyond the floor. This is done so that the coops may be tipped back, leaving the floor of the coop so that it may be easily cleaned. The sides and back of coops are made from store boxes and are not expensive. If painted they will last for years. This is the only kind of a brood coop used on my premises and the most satisfactory one I have seen to date.

MY EXPERIENCE IN THE POULTRY BUSINESS

U. R. Fishel, Hope, Ind., Tells of the Development of the Largest Specialty Poultry Farm in the World, Where More than Ten Thousand Thoroughbred White Rocks Are Bred and Sold Annually.

SOMETHING like thirty years ago the writer became interested in fancy poultry by reading a poultry paper published in Chicago and which was at that time one of the very few poultry publications. Having all my life had a liking for fowls and pets of any kind I naturally took to the poultry business readily. Being but a boy and with parents not blessed with an abundance of the world's goods it worried me how to secure a sitting of eggs from thoroughbred fowls. Fortunately I secured a job sawing wood with a buck saw, earning a dollar, with which I bought a sitting of thirteen Brown Leghorn eggs. Having labored hard for this money I was anxious to get the best results from the eggs. I will never forget the sight that met my eyes when I went to the nest on the nineteenth day. The trusty old hen was sitting on the nest with a fringe of little brown striped heads surrounding her. Twelve eggs of the thirteen had hatched. It is useless to say that I was delighted with my first venture in the poultry business. By caring for this brood carefully I succeeded in raising to maturity every chick, and there was but one cull in the lot, leaving me eleven good birds. This was my start in the poultry business.

By careful advertising I had no trouble in disposing of my surplus stock for the next few years. Like nearly everyone else that starts in the poultry business, I thought that I had to breed every variety there was, so I soon had something like fifteen varieties. One year's work along this line convinced me as it has almost everyone else that I was doing the most foolish thing possible. So I decided to specialize and keep but one breed. This I did and took up the Black Langshan, at that time a very popular fowl. We succeeded in winning gold medal sweepstakes prize at the New Orlean's World's Fair in 1889 for the best cock bird and refused for this bird the unheard of offer of $100. The following year we sold eggs at $8 per thirteen and had no trouble in disposing of every egg we had to spare. I mention these facts to show you that as far back as twenty years ago people were willing to pay good prices for fowls and eggs as long as they could get the quality.

After several years of successful business I decided that the poultry business was for boys and as foolish as it may seem I left the ranks of poultry fanciers and entered the mercantile business and although I succeeded in increasing the annual income of the business from $8,000 to $60,000 I could plainly see that there was more clean, clear money in the chicken business than in hardware and implements, so I gradually worked back into the poultry business, carrying it along with the mercantile business until I found someone who wanted a hardware store. The party was found and the deal consummated. I was then free to do as I pleased and I pleased to go into the poultry business right. I bought a 120-acre farm, nicely situated and well adapted to the poultry business. I decided to devote fifteen acres of this farm to poultry and stocked it with White Plymouth Rocks, which I have bred ever since. Buildings were erected and yards fenced—an expenditure of a couple of thousand dollars made. Neighbors said I was foolish and would loose every dollar I had made. I said nothing but kept on "sawing wood" and to-day I am pleased to say I have the entire 120 acres nearly covered with chickens. The Fishel White Rocks roam over the whole farm. We sell annually something like $40,000 worth of chickens and eggs. We ship over 10,000 White Plymouth Rocks each year, they going to all parts of the world—in fact, there is no place on the globe where chickens are known but that the Fishel White Rocks are talked of. From a hundred-dollar bird in 1887 we own to-day a thousand-dollar hen and we recently sold seven birds for $1,750. Which should prove to you, dear reader, that poultry will pay.

Some people say that there are too many failures in the poultry business. Why is this? Wherever you find a poultry plant that has failed you will find that the owner has spent his money and time in idleness or dissipation. Or has shown some palpable lack of foresight in starting his plant by expending too much money on houses and equipment and too little on stock or by making some equally serious mistake.

There is no business that will give you the pleasure and profit that the poultry business will and all that is necessary to make it a successful venture is to have in the first place a liking for the business and after that good common sense, a desire to work and above all manhood

U. R. FISHEL'S POULTRY FARM AT HOPE, IND. THE LARGEST SINGLE-VARIETY POULTRY FARM IN THE WORLD.

enough to leave "drink" alone. Start with the best birds possible and let people know through advertising that you have the quality of fowls and are attending to business, and I am sure that your investment will prove the most profitable one you have ever made. This is written by one who feels that he has demonstrated his knowledge of the business and who will welcome you at any time if you will visit "Fishelton," the largest specialty poultry farm in the world, where I believe I can prove to you that poultry pays.

SINGLE COMB WHITE LEGHORNS AND BUFF PLYMOUTH ROCKS.

White Leghorns, first cockerel and third pullet; Buff Rocks, first cockerel and first hen at Madison Square Garden, New York, January, 1905. Owned and exhibited by Orchard Hill Poultry Farm, 473 South Cleveland Avenue, St. Paul, Minn.

CAPONIZING AND ITS PROFITS

The Operation Not Difficult—Its Advantages—Losses in Operation—Profits and Markets.

By CLARENCE HEWES, Indianapolis, Ind.

CAPONIZING is not by any means a new practice, being as old probably as the Christian Era, although it is only recently that it has taken on any considerable commercial importance. Our European cousins being ever more alert than ourselves for appetizing edibles, have taken hold of the practice more quickly and more extensively, and the knowledge of the capon's superior flesh is known among them to every poulterer and marketman. The American poultrymen, particularly in the East, have taken up the practice in many instances with uniformly satisfactory results, and from them it is gradually spreading throughout the West and even our ever-careful farmers and small poultrymen are taking it up.

—The Operation is not Difficult.—

The operation is not at all difficult, even inexperienced persons are successful with their first operations, although their losses are of course larger than those of experienced operators; the danger is not so much in the possible fatality of the operation as in the possibility of incompleteness thereby producing slips, the beginner should take his time with his first experiment, and if he feels that he is inflicting too much punishment upon the fowl he should kill it and proceed with the operation upon the dead bird. Make yourself thoroughly familiar with the operation on this first fowl and proceed the same thereafter with live ones. If at any time you think that you have fatally injured a bird kill it at once and bleed it and it will be as fit for the table as any other fowl.

—The Advantages of Caponizing.—

The impression seems to have taken hold with many that the greatest advantage of the capon over the cockerel is in its superior size. In truth the capon is only a trifle larger than a well matured cockerel—that is a well cared for cockerel, but when a large number of uncastrated males are carried through the winter and all housed together they are in fact much inferior to the capon in weight as in every other respect. However, the real advantages of caponizing lie in the superior flesh that the capon carries and in his docility and tractability. Fifty to one hundred capons may be housed together with as much safety as so many hens, but that number of cockerels would be in a perpetual state of turmoil. Any one that has tried to carry fifty or more cockerels through the winter will understand this condition.

The flesh of the capon is of superior grain and flavor and in addition is tender. The market prices range from 12 to 18 cents per pound and all dealers that I have questioned are willing to pay these prices.

The great advantage of caponizing on the farm is that it offers a profitable means of disposing of the late hatched cockerels. Almost all chicks raised on the farm are disposed of as springers, for the early hatches this is a very profitable method, but for the later one it is not. I refer to the late May and June hatches. Such chicks reach the market when springers are plentiful and cheap. It would prove much more profitable to the farmer to keep these cockerels and caponize them.

The capon is simply an unsexed cockerel and the operation has just the same effect on the cockerel that castration has on a bull or boar—tames the disposition and permits the fowl to make growth without the flesh becoming tough or stringy.

—Losses in the Operation.—

With experienced operators the losses attendant upon the operation are about five per cent. This includes all that die from the operation either immediately or afterwards. The percentage of loss grows larger as one's experience is less. Some beginners lose as high as fifty per cent. of their first batch. Such losses as this are due to bunglesomeness and if such a person doesn't find his efficiency increasing rapidly he had best induce some one else to do his caponizing for him.

—Profits and Markets.—

A yearling cockerel when sold on the market brings ten to fifteen cents, some times as much as twenty-five. A capon of the same age will bring from $1.20 to $1.75. The food cost has been the same and the capon has undoubtedly been the less trouble. The above are market prices and can be obtained in any of our market centers. Our villages and towns do not make good markets for capons and one would have trouble in realizing market quotations there. In our larger cities, however, the market has never been even half way supplied, and I have questioned the fashionable butchers of Indianapolis with regard to the prices that they can pay and their ability to dispose of the trade. Their replies are that they will pay market prices and that they have never yet had enough capons to supply the demand, and that they feel confident of their ability to sell all shipments sent to them.

—Not a Cruel Practice.—

No doubt many persons who realize the advantages of caponizing are deterred from practicing the operation out of tender-heartedness, disliking to inflict pain on any dumb creature. In fact, the practice is far from being cruel and is not extremely painful. It is seldom indeed that the fowl gives voice to any outcry during the operation and I have many times seen the birds peck at the flies that alighted near them. The most pain seems to be caused the fowl by the incision and after that the bird seldom shows any evidence of pain whatever. When released the fowls will eat greedily from their long fast that preceded the operation.

It must be remembered, too, that the birds escape much suffering through the operation. Their pugnacious dispositions are changed and that period of nagging and fighting attended with bloody combs and sometimes blinded eyes that all cockerels experience, is escaped by the capon on account of its altered nature.

—The Instruments.—

Caponizing instruments are manufactured by several companies, but the different sets do not vary much either in style or price. The description given herewith includes all necessary instruments. All sets include the following: A knife for making the incision, a spreader for holding the ribs apart and the incision open, a hook for tearing away the inside membrane, a probe to push aside the entrails should they get in the way, an instrument for catching and removing the testicles and a pair of forceps.

The knife is merely a piece of well tempered steel capable of being brought to a very sharp edge. The spring spreader is a steel spring so constructed that it may be inserted in the incision and will hold itself there and keep the incision open so that the operator may see what he is doing. The small steel hook is for the purpose of tearing away the thin membrane that appears as soon as the incision is made. The instrument for removing the testicles varies in different sets, perhaps the most used instrument is the canula. This is a hollow steel tube arranged for the insertion of a fine wire, which is so placed that its ends project from the large end of the tube and the middle of the wire forms a loop at the other end, the end conveying the wire loop is thrust into the incision and the loop is thrown around the testicle, then the wire loop is tightened by drawing on the end of the wire protruding from the free end of the tube, and the testicle is removed by twisting the tube.

Some sets of instruments instead of having the canula are supplied with the slotted scoop, an instrument with a small slotted spoon-like blade. The edges of the slot are sharpened and the spoon is slipped under the testicle in such a way that the slot catches the cord to which the testicle is fastened and cuts it, leaving the testicle in the spoon to be removed.

Aside from the above instruments it is necessary to have a basin of water which should contain a portion of carbolic acid for antiseptic purposes and some bits of sponge to absorb any blood that may be spilt.

—The Caponizing Table.—

The operator should be supplied with some kind of stand to place the fowl upon. Many operators use a barrel. This does very well if the bottom is supplied with holes to permit of the use of cords and weights to hold the fowl's

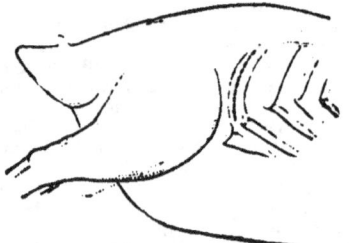

Fig. 1. The dotted line shows where incision should be made.

Spring spreaders for holding incision open.

feet and legs. It is best, however, to construct a table for the purpose on the following lines: It should be somewhat higher than the waist so as not to require much stooping. The top should be about two feet in diameter and should be so fastened to the stand that it may be tilted. This makes it possible to get the sunlight where you want it. The stand should also be fitted with holes through which the cords holding the fowl's wings and feet are placed. A box or bench should be placed near the stand to hold the instruments and other paraphernalia. The cords and weights referred to are cords with weights of about a pound each tied to each end. These are passed through openings in the stand and then over the fowl's legs or wings and the weights then hold them down and prevent any kicking or flopping while the bird is on the table.

—The Best Breeds—Age and Size.—

The American and Asiatic breeds make the best capons. The Orpingtons I think would also prove excellent, especially in markets demanding a white-skinned fowl. The operation should be performed just before the combs of the cockerels begin to "shoot." At this period the birds are

The knife for making the incision.

about three months old and weigh from one and a half to two pounds. Never caponize birds over five months old or weighing more than three pounds, as the operation would result fatally with so many that it would certainly prove unprofitable. Leghorns do not make good capons. They reach the age of caponizing before they are large enough to stand the operation and even when successfully operated upon they do not reach the weights desired in our markets.

—The Operation.—

The cockerels that you intend to caponize should be caught up and confined without food for at least twenty-four hours before the operation. This empties the bowels and causes them to be much less in the way. If the day set proves dark or cloudy postpone operations for a day. Bright sunlight is absolutely essential to enable the operator to see the interior of the fowl and the danger of any germ infection is much less on a bright day than on a dark, cloudy one.

If the day proves bright and clear get out your instruments and table and arrange them so as to be handy and catch your first fowl. Place the bird on the table on its side. Locate the last two ribs by feeling with the fingers and find the place pretty well up towards the back as shown by the dotted line in Figure 1. Pull out any feathers that are in the way and wet the surrounding ones and paste them to one side. Pull the loose skin to one side away from the point for the incision. Now take the knife for the initial incision, hold the blade upright and stick the point between the ribs and through the flesh and then pull it down, keeping it between the ribs all the time until you have an incision about one inch long. Care must be taken not to thrust the knife in so far as to endanger the bowels. Take the spring spreaders and compress them and insert them in the cut and release them, making sure that they are firmly fixed and will not fly out at some critical moment. You can now see a thin, filmy membrane covering the intestines. This must be torn away with the hook. The bowels must be carefully avoided in the meantime, as an injury to them would prove fatal. Now if your incision has been made in the right place the testicles can be seen almost directly under the opening, being fastened up close to the back. I believe that the making of the incision is the real test of the success of the operation, as when it is in the right place the testicles can be readily found, but when the incision is made too low down or too far forward it is very difficult, if not impossible, to locate them and proportionately harder to remove them.

When the testicles are seen take your canula, adjust the wire loop and insert the loop carrying end into the incision and slip the loop around the testicle onto the at-

The steel hook for tearing away the inside membrane.

taching cord. Now tighten the loop by drawing on the ends of the wire protruding from the free end of the tube. Sometimes the cord can be severed by merely pulling on the wire, but this is not often. Do not try to break the cord by pulling with the canula, but twist it slowly from side to side, and after several motions it will come loose and may be removed. If it should fall from the canula when severed it can be removed with the forceps. If your set contains the slotted scoop instead of the canula

The Canula, for removing the testicles. Note the arrangement of the wire.

the method of procedure is just the same except in the removal of the testicles. With the scoop the testicle is slipped into the scoop in such a way that the cord is caught in the slot. If the edges of the slot are well sharpened the cord will be cut leaving the testicle to be removed with the scoop. If, however, the edges are not sharp it will require some twisting of the scoop before the cord parts and allows of the removal of the testicle.

The operation given thus far is for the removal of one testicle from each side requiring two operations to capon-

The Slotted Scoop, preferred by many to the Canula.

ize the fowl. After having removed one testicle as above it is only necessary to turn the fowl and repeat the process on the other side. Many operators, however, remove both testicles from one side, removing the lower one first. This is a slightly more difficult operation than the former, but as it requires only one incision it is easier on the fowl and with experienced operators is a more rapid process.

The beginner had best content himself with making two incisions to the operations, and if during any operation he should find a specimen in which both testicles are in plain view from one side he may try the experiment of removing both from one side. If at any time during the operation any blood is shed it should be absorbed with pieces of sponge, first wetted in water containing some antiseptic and then squeezed dry. A small amount of blood spilled in the abdominal cavity is not liable to cause any subsequent trouble.

After having removed the testicles make sure that there are no foreign particles in the abdominal cavity; then remove the spreader and permit the skin to slip over the incision, which it will do if it has been properly pulled aside before the operation. The capon should be put in a quiet place for a few days and should be given food and water immediately. He will eat greedily from the long fast that preceded the operation. In a few days he may be released and should be treated much like any other growing chicken.

—Slips.—

Slips are the result of an incomplete operations where the testicles have not been completely removed. If the least portion is left it grows again and although the fowl never possesses any value as a breeder he causes almost as much trouble about the yards as uncaponized cockerels and possesses no more value when sold. Slips are avoided by carefulness in the operation. Make sure that the entire testicle is removed and you will have no slips.

—Marketing.—

In localities where collectors make frequent trips and are willing to pay reasonable prices it will undoubtedly prove best for the poultryman to sell his capons alive, thus escaping the responsibilities of dressing and shipping. It is not always possible to sell alive, however, as when one desires to sell to private or retail trade. Dry picking is the favored method for capons and in picking the feathers are left on the head and neck, on the wings from the "elbow" out and half way up the legs, the large stiff feathers of the tail are also left. This style of picking is considered characteristic of a capon.

DISEASES OF POULTRY

Prevention of Disease Lies in Care and Cleanliness—The More Common Diseases and How to Cure Them.

MOST of the diseases of poultry are caused by neglect or filth. This may be disputed by many, but nevertheless it is true, as I have been very careless at times. The great advancement which has been made by scientists in the study and investigation of the cause, origin and source of disease, the discovery of the specific germs, microbes and bacteria which are constantly associated with these disorders, the fact that an animal innoculated with fluid containing the germs or microbes identified with a disease from which another animal is suffering and that the disease is reproduced in the inoculated animal, the well known contagiousness of certain diseases, and their transmission from one animal to another, either by contact, contamination of food, drinking water or the atmosphere, can only be explained by what is now a question of dispute, the presence of a specific organism, or microbe, the causation poisonous agent of that particular disease, conveyed and communicated to each other by means of the agencies just mentioned. In the light of this knowledge the "germ theory of disease" applies to the lower orders of animals as well as the human, and a proper appreciation of the means and methods of preventing the development or afterward destroying these germs of disease is of paramount importance to the breeder of poultry, for quite often the prompt isolation of an affected bird, and thorough and vigorous disinfection of all the surroundings, will prevent a spread of the disease and save his flock from total annihilation.

There are certain agencies which produce the death of these germs, render them inert and powerless to develop disase. These are sunshine, fresh air, perfect cleanliness, the free use of certain chemical compounds, as carbolic acid and its modifications, bichloride of mercury, boiling water, etc. These agencies are what are called germicidal or germ-killing antiseptic and disinfectant. They produce the death of the germ or microbe. There are many good disinfectants on the market, and by using any one of these freely a great deal of disease may be avoided.

There are cases, however, occurring among the healthiest birds, where exposure to wet or cold will bring about an acute attack of disease, the symptoms of which are perfectly familiar, and where prompt and appropriate treatment undertaken immediately will in the majority of cases effect a perfect cure. Cases of this kind that are well understood and where the proper treatment is known are not difficult of treatment, but on the other hand quite frequently a fowl seems to be ill, and there are no positive or definite sysmptoms to enlighten you as to the true nature of the disorder nor any special indications for a fixed line of treatment. In these cases it is probably better to give the bird good, careful nursing, sustaining food, watching carefully the conditions, being prepared to meet any unusual emergency and trusting to natural powers of recuperation to produce a cure.

The principal diseases to which poultry are subject and the appropriate treatment I will briefly describe.

Bumble Foot.

This name is given to the affection consisting of corn or abscess at the bottom of the foot, and is more prevalent among the heavy-weight varieties than among the smaller ones.

Treatment.—If taken in hand early and the foot washed several times daily in a solution of carbolic acid water and afterward touching with lunar caustic or painting with tincture of iodine the diseased surface, a cure may be often affected. If let run on until the tumor becomes enlarged, or where pus has formed, it becomes necessary to open the swelling and let the matter or hard cheesy substance out, after this wash out the wound twice a day with one to fifty solution of carbolic acid water, keep the bird from roosting on a perch until foot is well.

Catarrh.

This disorder is the common cold to which all fowls are subject, and is manifested by a watery discharge from eyes or nostrils, general lassitude and weakness, and if neglected may develop into roup.

Treatment.—Keep the fowl in a warm, dry place, and place five drops of tincture of aconite in half a pint of drinking water. Feed moderately on soft food, mixed warm, and seasoned with a portion of the following mixture: Iodine, 2 oz.; ginger, 2 oz.; aniseed, 1-2 oz.; pimento, 2 oz.; cayenne pepper, 1 oz., and sulphate of iron, 1 oz. Rub and mix well together. This should produce a cure within a few days. If, however, the fowl does not improve, but seems to grow worse, treat for roup, which has probably supervened.

Cholera.

This disease is epidemic in character, highly contagious, due to a specific microbe or germ which communicates the disease to the other fowls. The disease is characterized by sudden thirst, excessive diarrhoeal discharges, first of a greenish color, afterward becoming white and watery, like the characteristic "rice water" discharges in the human subject.

Cause.—The active exciting cause is the virulent infectious cholera microbe or germ. The secondary or contributing causes are generally undue exposure to excessive heat of the sun without ample shade and warm drinking water. These, together with accumulation of filth and unhealthy surroundings, develop the cholera "bacillus," or "microbe," which is the primary cause of the disease. Filth breeds disease and is especially favorable soil for cholera germs, without the destruction of which the disease cannot be eradicated. The disease spreads among the other fowls by the offensive droppings contaminating the grass, drinking water, etc.

Symptoms.—Sudden and violent thirst accompanied with diarrhoea, great weakness and falling about, a peculiar "anxious" look about the face.

Treatment.—This being a germ disease, the successful way of combating it is to use a disinfectant which destroys the germs and renders the atmosphere pure and sweet. Sprinkle the houses, coops and runs with "crude carbolic acid;" keep birds in cool, airy places, with plenty of fresh air and abundance of shade and cool drinking water. Give plenty of fresh, green food. The disease is very fatal, death occurring in from twelve to forty-eight hours. If discovered early enough, about seventy-five per cent. of the birds may be saved by administering every three hours rhubarb, five grains; cayenne, two grains; and laudanum, ten drops. Also give between doses a teaspoonful of brandy diluted with a little water to which has been added two drops of carbolic acid. Five drops of carbolic acid should be added to each quart of the drinking water in order that the other birds may escape.

Crop Bound.

The crop is distended with hard grain, afterward swelling by the secretions, causing the outlet to be closed by the pressure.

Treatment.—Pour warm water down the throat, quietly and patiently knead the crop for an hour or more. Though it may be hard at first, it will yield after a time and become soft. When relaxed, hold the fowl with head down and work the crop, which will force the grain out of the mouth; pour more warm water down throat and repeat several times. This will force quite a little out; after which give a teaspoonful of castor oil and place the bird in a coop and give no food for twelve hours and feed sparingly for a few days in order to allow the organ to contract, or permanent distension may ensue. If these measures fail (which seldom do unless gone for quite a while without treatment) make an incision one inch long near the top of a crop, avoiding any blood vessel, and remove the contents of the crop, taking care that no hard substance is left behind; wash out well with warm water, slightly carbolized, sew up the inner membrane with fine silk thread, making three or four separate stitches. In like manner sew up the outer skin, feed on moist food and give no water for twenty-four hours.

Crop Soft or Swelled.

The contents are soft or fluid. The disorder being due to a lack of tone of the inner coats, resulting in inability to contract on the food.

Treatment.—Place the bird alone, and give small portion of cooked food three times a day, and a small quantity of water to be given after each meal only, the water to be slightly acidulated with nitric acid. Constant care in treatment is necessary to success.

Debility.

Cause.—Strain on nervous system, severe shock, excessive terror.

Symptoms.—Drooping with no apparent disease, lack of tone, prostration.

Treatment.—Judicious care, nourishing food, put in a quiet place and treat gently so as not to get bird excited. Give a tonic for a few weeks, consisting of iron and quinine mixture.

Diarrhoea.

It is a very common complaint.

Cause.—Sudden change in diet or weather.

Treatment.—In the beginning it may be checked by giving two or three meals of well boiled rice dredged with flour or powdered chalk. A diet of boiled oat meal or barley meal is very effectual.

Dysentery.

A contagious disease characterized by an inflammation of the lower bowels, with bloody stools and straining. It is rarely cured.

Treatment.—Complete and thorough disinfection, with a few drops of carbolic acid in drinking water. Restrict diet, give laudanum five drops, carbolic acid two drops in a teaspoonful of diluted whiskey every three hours.

Egg Bound.

Inability to lay on account of unusual size of eggs.

Symptoms.—Hen comes off nest without laying, walks slowly around with wings hanging down, and appears in distress.

Treatment.—Give dose of castor oil and handful of groundsel. This often gives relief. If not, inject an ounce of olive oil by means of a small flexible syringe and apply a one-to-one-hundred solution of carbolic acid and water to the vent to relax the tissues. Handle the bird gently so as not to break the egg, as this would mean sure death. Sometimes the passage enclosing the egg presents externally, ruptures, and the egg passes through. In such case egg-production must be stopped by giving every four hours a pill of calomel, one grain; tartar emetic, 1½ grain, and opium, ¼ grain. Put the bird on soft unstimulating diet.

Elephantiasis, or Scaly Egg.

Rough, scrofulous, unsightly scurfs on shanks, most common in Asiatic breeds.

Cause.—A small insect which collects under the scales on shanks, will be found on birds that stay about barn or manure pile mostly.

Treatment.—Can generally be cured. Place fowl upon floored run, wash feet and legs with a nail brush and apply an ointment consisting of sulphur and lard with enough kerosene to mix well. Apply daily, rubbing under scales well; also give small dose of sulphur in mash food.

Feather Eating.

This affection consists in devouring of each others' plumage, picking the feathers until the blood flows often. This is most common in the French and Polish breeds.

Cause.—Idleness, being confined to close quarters with nothing to keep them busy. It often occurs among exhibition birds while at the shows, and the only way to stop it in this case is to coop separately, or place a bit in the mouth made of small wire, extending from nose holes to nose holes, passing under the upper bill, which holds the mouth open sufficiently far that they cannot hold a feather, yet does not interfere with their eating.

Frost Bites.

The parts generally affected are the comb, wattles and sometimes toes.

Treatment.—Rug vigorously with snow or hold in ice water and rub until frost is all out. Afterward apply a mixture of glycerine, two parts; kerosene, one part, three times daily, and keep bird in warm place. This accident can generally be prevented by oiling the parts each morning and evening. The sure way is a good, warm house.

Gapes.

This is a disease of chickens or young fowls and consists in the wind-pipe becoming infested with small, white worms, causing the chicken to strangle and gape for breath, waste away, and finally die from actual suffocation. The worm is about three-fourths of an inch in length, of a pale, reddish color, and the number in each chick varies from two to twelve. It is most always found double, a smaller worm (the male) being attached about one-fourth from the upper end. These worms develop from an egg which finds its way into the chicken's mouth, caused from eating angle worms. Take a rich soil that is damp, and you will find more gapes as they get more angle worms.

Treatment.—Swab throat with a strong solution of 1-3 kerosene, 1-3 turpentine and 1-3 lard.

Indigestion.

Cause.—Injudicious use of special foods, unwholesome diet.

HOW TO MAKE POULTRY PAY.

Chief Cause.—Overfeeding, resulting in inflamed stomach, sluggish liver or simple debility.

Symptoms.—Bird walks lazily about, with no appetite and droppings scanty and unhealthy in color.

Treatment.—If digestive system is deranged, give daily: Rhubarb, 5 grains, changed every fourth day to calomel, one grain. Restrict diet to small portion of well cooked food, twice daily. A little water may be given after meals only.

Leg Weakness.

Frequent affection in cockerels of large breeds, due to outgrowing the strength of the legs, and occurs between ages of three and six months.

Cause.—Muscular weakness or bony deficiency.

Symptoms.—More or less squatting on the ground, instead of walking around. If the weakness is caused by a deficiency of bony matter, there is a tendency to knock knees or crooked breast bone.

Treatment.—Give bone dust in food freely, also a pill of sulphate of iron, 1 grain; strychnine sulphate, 1-30 grain; phosphate of lime, 5 grains, and quinine sulphate, 1-3 grain. Give three times daily. After a week or two give instead, half a teaspoonful of "Parrish's Chemical Food" in teaspoonful of water, morning and night.

Pip.

Horny appearance at end of tongue.

Cause.—May be due to obstruction of nostrils, causing bird to breathe through mouth. Drying end of tongue, or if due to a real affection is analagous to foul tongue in human.

Treatment.—Give cathartic; remove any crust or scale that will come off, and wash tongue and mouth with a mild solution of carbolized water.

Rheumatism.

Cause.—Exposure to cold or wet, privation, deficient nourishment.

Symptoms.—Weakness of legs, stiffness of the joints, or contraction of the toes. Distinguished from "leg weakness" by the latter affecting young birds, and due to muscular weakness.

Treatment.—Put fowl in warm, dry place; give stimulating diet. In evening put in the basket by the fire. Bathe legs in warm water to which some mustard has been added. Internally give daily a pill consisting of salol, 2 grains, and opium, one-half grain. Minute doses of oil of mustard in the food is serviceable. Regularity in treatment is essential.

SEVEN WHITE ROCKS THAT SOLD FOR $1,750.
Sold by U. R. Fishel, Hope, Ind., to William Miller, Crescent, Mo.

SELECTING A BREED

The Beginner Should Make His Choice Largely on His Own Judgment and Be Loyal to that Choice When Once 'Tis Made.

IN the following forty-eight pages of this book will be found illustrations of all standard and popular breeds of poultry. These should enable the beginner to easily and satisfactorily solve the first problem that confronts him, namely, the selection of a breed.

The many different breeds of poultry with their sub-varieties offer so many combinations of color, shape and characteristics that surely some can be found that appeal to the fancy of any individual.

The beginner should make his choice in this matter largely on his own judgment, for this is one of those cases where much advice only tends to confuse. Consult your own tastes and desires and under no circumstances take up with a breed unless it appeals to your ideas of the beautiful; unless you admire it for its peculiarities and differences as well as for its practical worth. Success with poultry depends so much upon personal care and attention that it is worth much to possess a flock of fowls that really appeal to one's affections, for the care of such a flock will be easy and spontaneous, where under different circumstances it would be tedious work and drudgery.

It is well to remember that many an excellent breed is to-day in the background and numbered among the less popular varieties merely for want of a few energetic breeders to make its merits known. Many of these varieties are quite equal in practical value to the more popular breeds. If you find your fancy running toward the more obscure breeds do not for that reason fear to trust your judgment, for it is inevitable that many of these meritorious fowls shall again find their way into popular favor. Do your part toward pushing your favorites forward to the place in the van to which you think they are entitled.

If your choice falls on any of the popular breeds, such as the Rocks or Wyandottes, all is well and good. You will find many sources from which to get your foundation stock or eggs and there will be many other advantages arising from the popularity of the fowl which you take up. But on the other hand you will find when you become a showman or advertiser that you also have many contestants.

As for new varieties there are few years that go by without seeing one or more added to our already long list of thoroughbred fowls, and in many instances they have a steady and rapid development and are soon admitted to the American Standard of Perfection. The progress of new breeds usually starts with a "boom" which proves profitable for those who are fortunate enough to be in when the "boom" starts. Then there is a reaction following this. If the breed is really one of merit there develops a normal demand. It is better for the beginner to let new breeds alone until after the "boom" period is over.

After having made a selection stand by it and give the breed every chance to demonstrate its worth. If you desire a reputation as a breeder of exhibition stock do not be disappointed that it does not come immediately. It requires several successes in the show room and a considerable expenditure for advertising before rival breeders will even recognize your presence in the fancy. Your first exhibits unless made in small or local shows will more likely be defeats than successes, but these are necessary as a part of every fancier's education.

Do not regard expenditures as over when you have secured your first stock, for no man can come to the front or stay there after once having gotten there unless he is a wise and judicious buyer as well as a careful and painstaking breeder. The poultry world has become so large that no one man can breed all the good ones, and even our best breeders are constantly on the lookout for birds that will help their flock in this or that particular, and when they find them they do not quibble about the price.

It is not absolutely essential that one be a specialist, although there is no doubt that it is much better for the beginner to take up with but one variety for the first few years of his apprenticeship to the fancy. It is a difficult thing for a real fancier to confine his affections to one breed of fowls and there are few who do not feel moved after a few years in the fancy when they have had opportunities to become acquainted with the beauties of many breeds to take up some more breeds. For the most part this is done merely to satisfy the innate affection for feathered beauty, but in spite of the oft-repeated statement that specialists are the most successful poultrymen the fact remains that there are many variety men who are making great successes. There are many breeders keeping many varieties who push only one of them to the front. The others they keep simply because they like them, and they evidently think that the pleasure they get out of them as fanciers is worth all they cost.

The purchase of the stock itself is a task indeed, for not one beginner in ten realizes that he is laying the foundation of a future enterprise with which he is liable to concern himself for the remainder of his days. It must be remembered that if the start is made with dollar birds that practically all of the stock raised from them will be likewise dollar birds; while if the beginning be made with high quality stock the very first breeding season will end with a flock of high class birds filling the fanciers' yards. It is far better that one's flock be small and good than that it be large and of indifferent quality. The greatest advantage, however, of a good start with high class birds is that it puts the breeder in a position to cope with breeders who have had several years the start of him, but who failed to take advantage of this same opportunity to start right. There are many fanciers, competent breeders and men who make the most of the stock that they possess who after years in the business are still far from the top for no other reason than that they are unable to appreciate the good that a few high class birds would do in their flock and hesitate at the high prices, though they should have learned with their experience that price is commensurate with quality and that true quality always commands good prices.

The selection of a breed is but the first step for the beginner. Success depends largely on the loyalty that the new fancier shows toward the breed of his choice and the energy and determination he possesses and exerts in bringing his birds to the highest stage of perfection.

CUTS FOR POULTRYMEN

The illustrations in the following pages of this book, aside from being a valuable guide to beginners in the selection of a breed, also constitute a Catalogue of Poultry Cuts. These cuts are made from original drawings by Mr. I. W. Burgess and conform to the requirements of the latest revised Standard of Perfection.

There is no question but that the value of advertising and printed matter is greatly increased by the liberal use of appropriate illustrations—they will convey a correct and certain impression where reading matter at best would be tiresome and cumbersome.

An electrotype of any illustration in the following forty-eight pages of this book will be sent postpaid on receipt of the price which will be found quoted under each illustration.

When ordering cuts give number of page on which the illustration appears, name of breed which the illustration represents, and the number of the cut which will be found immediately below it.

Address all orders to
The Inland Poultry Journal Publishing Co.,
Indianapolis, Ind.

Copyright by Inland Poultry Journal Company, Indianapolis, Ind.

SILVER SPANGLED HAMBURGS.

BY I. W. BURGESS.

An accurate, lifelike and honest reproduction of this most beautiful breed of fowls. In the above illustration Mr. Burgess has given us by far the best representation of this breed that has ever been put forth. We have more such treats as this in store for our readers.

www.ingramcontent.com/pod-product-compliance
Lightning Source LLC
Chambersburg PA
CBHW082358220526
45470CB00008B/2790